LONDON MATHEMATICAL SOCIETY LECTURE NOTE SERIES

Editor: PROFESSOR G. C. SHEPHARD, University of East Anglia

This series publishes the records of lectures and seminars on advanced topics in mathematics held at universities throughout the world. For the most part, these are at postgraduate level either presenting new material or describing older material in a new way. Exceptionally, topics at the undergraduate level may be published if the treatment is sufficiently original.

Prospective authors should contact the editor in the first instance.

Already published in this series

1. General cohomology theory and K-theory, PETER HILTON.
2. Numerical ranges of operators on normed spaces and of elements of normed algebras, F. F. BONSALL and J. DUNCAN.
3. Convex polytopes and the upper bound conjecture, P. McMULLEN and G. C. SHEPHARD.
4. Algebraic topology: A student's guide, J. F. ADAMS.
5. Commutative algebra, J. T. KNIGHT.
6. Finite groups of automorphisms, NORMAN BIGGS.
7. Introduction to combinatory logic, J. R. HINDLEY, B. LERCHER and J. P. SELDIN.
8. Integration and harmonic analysis on compact groups, R. E. EDWARDS.
9. Elliptic functions and elliptic curves, PATRICK DU VAL.
10. Numerical ranges II, F. F. BONSALL and J. DUNCAN.
11. New developments in topology, G. SEGAL (ed.).
12. Proceedings of the Symposium in complex analysis, Canterbury 1973, J. CLUNIE and W. K. HAYMAN (eds.).
13. Combinatorics, Proceedings of the British combinatorial conference, 1973, T. P. McDONOUGH and V. C. MAVRON (eds.).
14. Analytic theory of abelian varieties, H. P. F. SWINNERTON-DYER.
15. Introduction to topological groups, P. J. HIGGINS.
16. Topics in finite groups, TERENCE M. GAGEN.
17. Differentiable germs and catastrophes, THEODOR BRÖCKER and L. C. LANDER.
18. A geometric approach to homology theory, S. BUONCRISTIANO, C. P. ROURKE and B. J. SANDERSON.
19. Graph theory, designs and coding theory, P. J. CAMERON and J. H. VAN LINT.

London Mathematical Society Lecture Note Series. 20

Sheaf Theory

B.R. TENNISON

CAMBRIDGE UNIVERSITY PRESS
CAMBRIDGE
LONDON · NEW YORK · MELBOURNE

CAMBRIDGE UNIVERSITY PRESS
Cambridge, New York, Melbourne, Madrid, Cape Town, Singapore, São Paulo

Cambridge University Press
The Edinburgh Building, Cambridge CB2 8RU, UK

Published in the United States of America by Cambridge University Press, New York

www.cambridge.org
Information on this title: www.cambridge.org/9780521207843

First published 1975
Re-issued in this digitally printed version 2007

A catalogue record for this publication is available from the British Library

Library of Congress Catalogue Card Number: 74-31804

ISBN 978-0-521-20784-3 paperback

Dedicated to my family

Contents

Introduction

Sheaf theory provides a language for the discussion of geometric objects of many different kinds. At present it finds its main applications in topology and (more especially) in modern algebraic geometry, where it has been used with great success as a tool in the solution of several long-standing problems. In this course we build enough of the foundations of sheaf theory to give a broad definition of manifold, covering as special cases the algebraic geometer's schemes as well as the topological, differentiable and analytic kinds; and to define sheaf cohomology for application to such objects.

Chapters 1 and 2 cover the groundwork of presheaves and sheaves, and show that any presheaf gives rise to a sheaf in a universal way.

Chapter 3 defines the categorical viewpoint, shows that the categories of sheaves and presheaves of abelian groups on a fixed topological space are abelian, and investigates the relations between them. It also covers the processes of change of base space of a sheaf, both for the inclusion of a subspace and for a general continuous map.

Chapter 4 defines the notions of ringed space and geometric space, and gives as an example the spectrum of a commutative ring. This is proved to be a construction with a universal property among all geometric spaces. The latter are shown to be the prototypical geometric objects, by exhibiting various kinds of manifolds as special cases. The chapter also includes a discussion of Modules over ringed spaces, and the consideration of locally free Modules leads to the definition of the picard group of a ringed space.

Chapter 5 gives an introduction to sheaf cohomology, at first in the general context of being the right derived functor of a suitable left exact functor between abelian categories. It is shown how other definitions fit into this picture, particularly those defined by flasque sheaves and by the Čech method. The picard group of a ringed space is interpreted as a cohomology group.

The essence of the usefulness of sheaves is that they express the connexions between the local and global properties of a geometric object. This should become clear from the results of Chapters 4 and 5.

The approach to the subject taken here is rather categorical, and the course may be used (and indeed has been, in Part III of the Mathematical Tripos at Cambridge) as an introduction to the usefulness of categories and functors. It presupposes only a knowledge of elementary general topology (topological spaces and open sets) and elementary algebra (abelian groups, rings), although reference is made to other sources for further elucidation of some points.

There are exercises scattered throughout the text and at the end of each chapter, and they vary considerably in difficulty. There is a section at the end of the book containing hints and solutions to some of them.

Conventions and notation

We use Bourbaki notation for the sets **N**, **Z**, **Q**, **R**, **C** of natural numbers (0, 1, 2, ...), integers, rationals, reals and complexes. We use the barred arrow $\mapsto$ to indicate where an element of a set is sent under a map with that set as domain: hence for example

$$f : \mathbf{Z} \to \mathbf{Z} : n \mapsto n^2$$

defines f as the squaring map. $\cong$ and $\xrightarrow{\sim}$ each denote an isomorphism. $\amalg$ denotes disjoint union of sets, definable for instance as

$$\amalg_{\lambda \in \Lambda} X_\lambda = \cup_{\lambda \in \Lambda} \{\lambda\} \times X_\lambda;$$

the important thing is that $\amalg_{\lambda \in \Lambda} X_\lambda$ is the disjoint union of a copy of X_λ for each λ.

References within the book are explained by example as follows. 4.3.12 is the full reference to part 12 of §3 of Chapter 4; within Chapter 4 this is abbreviated to 3.12. 4.Ex.3 refers to Exercise 3 at the end of Chapter 4.

References to other sources are either quoted in full or given as [X], where X is one of a short list of acronyms detailed in the list of references (page 156).

1· Presheaves and their stalks

In this chapter, we give definitions and examples of presheaves of sets and of abelian groups, and of morphisms between them. We study the notion of direct limit of a directed system of sets (or abelian groups), and apply it to construct the stalks of a presheaf, which summarise the nature of the presheaf locally in the neighbourhood of some point.

1.1 Definition of presheaves

1.1 **Definition.** Let X be a topological space. A presheaf F of sets on X is given by two pieces of information:

(a) for each open set U of X, a set $F(U)$ (called the set of sections of F over U)

(b) for each pair of open sets $V \subseteq U$ of X, a restriction map $\rho^U_V : F(U) \to F(V)$ such that

(b1) for all U $\rho^U_U = id_U$

(b2) whenever $W \subseteq V \subseteq U$ (all open) $\rho^U_W = \rho^V_W \circ \rho^U_V$ i.e.

$$\begin{array}{ccc} F(U) & \longrightarrow & F(W) \\ & \searrow \quad \nearrow & \\ & F(V) & \end{array}$$

commutes.

A presheaf of abelian groups over X is a presheaf F of sets such that

(a') each $F(U)$ has a given abelian group structure

(b') every restriction map ρ^U_V is a group homomorphism with respect to these structures.

1.2 **Remark.** Sets and abelian groups are the two main types of structure with which we shall be concerned for the moment; it should be clear how to phrase the definitions of presheaves of groups, rings, commutative rings, topological spaces,...; all the sets of sections have the appropriate structure, and all the restriction maps are morphisms of the appropriate kind (homomorphisms, continuous maps, ...).

1.2 Examples of presheaves

Example A. Let A be any given set (or abelian group). Then the constant presheaf A_X on X is given by

$$\begin{cases} A_X(U) = A & \text{for } U \text{ open in } X \\ \rho^U_V = \mathrm{id}_A : A_X(U) \to A_X(V) & \text{for } V \subseteq U \text{ open in } X. \end{cases}$$

Example B. Let Y be another topological space. The presheaf C^Y of continuous Y-valued functions on X is defined by:

$$\begin{cases} C^Y(U) = \text{set of continuous maps} : U \to Y & (\text{for } U \text{ open in } X) \\ \rho^U_V : C^Y(U) \to C^Y(V) : f \mapsto f|V & (\text{for } U \supseteq V \text{ open in } X) \end{cases}$$

[whence the name 'restriction' for the maps ρ^U_V].

If in addition Y has the structure of an abelian group, so has each $C^Y(U)$ by pointwise addition of functions. In this case C^Y is a presheaf of abelian groups. For example, giving $\mathbf{Z}$ the indiscrete topology, $C^{\mathbf{Z}}$ is a presheaf of abelian groups on X [it is even a presheaf of rings]. Similarly for $C^{\mathbf{R}}$ for $\mathbf{R}$ with the usual topology.

Example C. Suppose that X is an open subset of some $\mathbf{R}^n$. Let $r \in \mathbf{N}$ (= {0, 1, 2, ... }). The presheaf C^r of r-times differentiable $\mathbf{R}$-valued functions on X has

$$C^r(U) = \text{set of r-times continuously differentiable functions} : U \to \mathbf{R}$$
(for open U)

and restrictions as in Example B.

Example D. Suppose that X is an open subset of some $\mathbf{C}^n$. The presheaf C^ω of analytic $\mathbf{C}$-valued functions on X has

$$C^\omega(U) = \text{set of analytic functions} : U \to \mathbf{C} \quad (\text{for } U \text{ open in } X)$$
(analytic = regular = holomorphic).

Example E. Two more pathological examples. Let X be any topological space with more than one point e.g. $X = \{0, 1\}$ or $[0, 1] \hookrightarrow \mathbf{R}$.

Define the presheaf P_1 by

$$\begin{cases} P_1(X) = \mathbf{Z} \\ P_1(U) = \{0\} \text{ (trivial group) for open } U \neq X \\ \text{all restrictions except } \rho^X_X \text{ being constant maps.} \end{cases}$$

Pick $x_0 \in X$. Define the presheaf P_2 by

$$\begin{cases} P_2(U) = \mathbf{Z} \text{ for } U \text{ open in } X \text{ such that } U \ni x_0 \\ P_2(U) = \{0\} \text{ for } U \text{ open in } X \text{ such that } U \not\ni x_0 \\ \text{restrictions } \rho^U_V = \begin{cases} \mathrm{id}_{\mathbf{Z}} \text{ if } x_0 \in V \subseteq U \\ \text{trivial map if not.} \end{cases} \end{cases}$$

Then P_1, P_2 are both presheaves of abelian groups on X.

1.3 Interlude: direct limits

3.1 **Definition.** A directed set Λ is a set with a pre-order $\leq$ (that is, a reflexive and transitive relation: $\alpha \leq \alpha$, and $\alpha \leq \beta \leq \gamma \Rightarrow \alpha \leq \gamma$) which also satisfies:

(a) $\forall \alpha, \beta \in \Lambda$ $\exists \gamma \in \Lambda$ such that $\alpha \leq \gamma$ and $\beta \leq \gamma$.

We often write $\Lambda_1 = \{(\alpha, \beta) \in \Lambda \times \Lambda;\ \alpha \leq \beta\}$.

A direct system of sets indexed by a directed set Λ is a family $(U_\alpha)_{\alpha \in \Lambda}$ of sets together with, for each $(\alpha, \beta) \in \Lambda_1$, a map of sets $\rho_{\alpha\beta} : U_\alpha \to U_\beta$, satisfying

(b) $\forall \alpha \in \Lambda$ $\rho_{\alpha\alpha} = \mathrm{id}_{U_\alpha}$

(c) $\forall \alpha, \beta, \gamma \in \Lambda$ if $\alpha \leq \beta \leq \gamma$ then the triangle

$$\begin{array}{ccc} U_\alpha & \longrightarrow & U_\gamma \\ & \searrow \quad \nearrow & \\ & U_\beta & \end{array}$$

commutes, i.e. $\rho_{\alpha\gamma} = \rho_{\beta\gamma} \circ \rho_{\alpha\beta}$.

3.2 **Example.** Given a topological space X, the set T of its open sets is directed by the relation

$$U \le V \iff U \supseteq V$$

(condition (a) holds since $U \cap V$ is open if U, V are).

Given also a presheaf F on X, let $\rho_{UV} = \rho^U_V$ be the restriction map when $U \le V$. Then the $(F(U))_{U \in T}$ with the ρ_{UV} form a direct system of sets.

3.3 Picture.

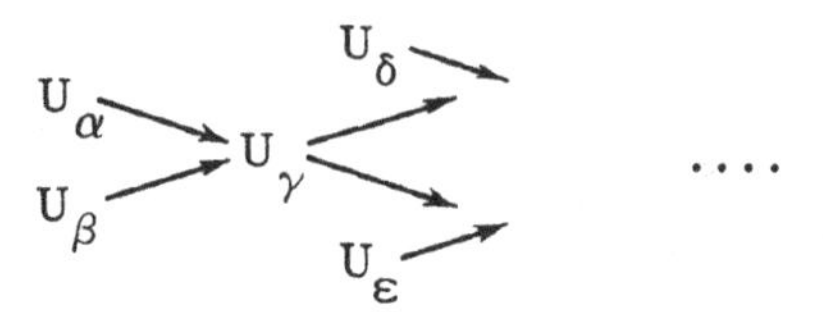

....

We wish to find a 'limit' for the system, i.e. an object which can go 'to the right of everything'.

3.4 Definition. Given a direct system in the notation of 3.1, a <u>target</u> for the system is a set V and a collection of maps $(\sigma_\alpha : U_\alpha \to V)_{\alpha \in \Lambda}$ satisfying the compatibility condition:

$$\forall \alpha \le \beta \qquad \begin{array}{ccc} U_\alpha & \xrightarrow{\sigma_\alpha} & V \\ {\scriptstyle \rho_{\alpha\beta}} \downarrow & \nearrow {\scriptstyle \sigma_\beta} & \\ U_\beta & & \end{array}$$

commutes; that is $\sigma_\alpha = \sigma_\beta \circ \rho_{\alpha\beta}$.

A <u>direct limit</u> for the system is a target U, $(\tau_\alpha : U_\alpha \to U)_{\alpha \in \Lambda}$ satisfying the universal property:

(*34) for any target V (with maps σ_α as above) there is a <u>unique</u> map $f : U \to V$ such that

$$\forall \alpha \in \Lambda \qquad \begin{array}{ccc} U_\alpha & \xrightarrow{\sigma_\alpha} & V \\ {\scriptstyle \tau_\alpha} \searrow & & \nearrow {\scriptstyle f} \\ & U & \end{array}$$

commutes.

3.5 Remark. So a direct limit is a 'best' target.

3.6 **Proposition.** Any two direct limits for a direct system are naturally isomorphic (that is, there is a bijection between them compatible with all the τ_α).

Proof. (Archetypal of a large number of proofs of similar statements.) Let them be $(U, (\tau_\alpha)_{\alpha\in\Lambda})$ and $(U', (\tau'_\alpha)_{\alpha\in\Lambda})$. Since U is universal and U' a target, we obtain an $f : U \to U'$; since U' is universal, we obtain a $g : U' \to U$ and

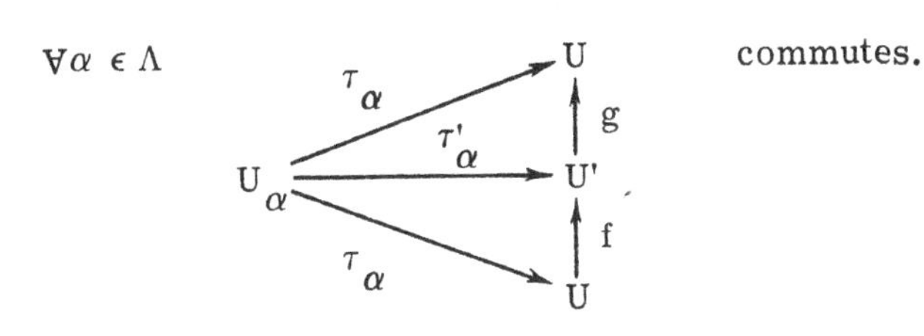

But now U is a target, and the universality of U implies that $id_U : U \to U$ is the unique map making all the

$$U_\alpha \xrightarrow{\tau_\alpha} U \uparrow, \quad U_\alpha \xrightarrow{\tau_\alpha} U$$

commute. Hence $g \circ f = id_U$ and similarly $f \circ g = id_{U'}$. //

3.7 **Notation.** Thus there is some justification in speaking of the direct limit and denoting it by $\varinjlim_{\alpha\in\Lambda} U_\alpha$.

3.8 **Theorem.** Suppose $U, (\tau_\alpha : U_\alpha \to U)_{\alpha\in\Lambda}$ is a target for the system $(U_\alpha)_{\alpha\in\Lambda}, (\rho_{\alpha\beta})_{(\alpha,\beta)\in\Lambda_1}$, such that:

(i) $\forall u \in U \;\exists \alpha \in \Lambda$ such that $u \in \mathrm{Im}(\tau_\alpha)$

(ii) if $\alpha, \beta \in \Lambda$ and $u_\alpha \in U_\alpha$ and $u_\beta \in U_\beta$ then

$$\tau_\alpha(u_\alpha) = \tau_\beta(u_\beta) \iff \exists \gamma \in \Lambda \text{ such that } \alpha \le \gamma,\ \beta \le \gamma \text{ and } \rho_{\alpha\gamma}(u_\alpha) = \rho_{\beta\gamma}(u_\beta).$$

Then U is a direct limit of the system.

[Remark: We can paraphrase (ii): if u_α, u_β get identified in U, they must have been identified along the way.]

Proof. Suppose $V, (\sigma_\alpha)_{\alpha \in \Lambda}$ is another target. If $f : U \to V$ is to satisfy the compatibility condition (*34) then it must be obtained as follows:

$$(*38) \quad \begin{cases} \text{for } u \in U, \text{ pick } \alpha \in \Lambda \text{ such that } u \in \mathrm{Im}(\tau_\alpha), \text{ say} \\ \qquad\qquad\qquad\qquad u = \tau_\alpha(u_\alpha); \\ \text{then } f(u) = \sigma_\alpha(u_\alpha). \end{cases}$$

Hence, if f exists, it is unique.

If we choose $\beta \in \Lambda$ such that $u \in \mathrm{Im}(\tau_\beta)$ too, say $u = \tau_\beta(u_\beta)$, then by condition (ii) $\exists \gamma \in \Lambda$ with $\rho_{\alpha\gamma}(u_\alpha) = \rho_{\beta\gamma}(u_\beta)$; hence

$$\begin{aligned} \sigma_\alpha(u_\alpha) &= \sigma_\gamma(\rho_{\alpha\gamma}(u_\alpha)) \\ &= \sigma_\gamma(\rho_{\beta\gamma}(u_\beta)) \\ &= \sigma_\beta(u_\beta). \end{aligned}$$

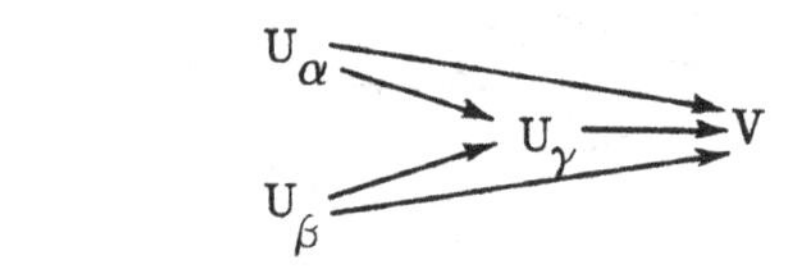

So f is well-defined by (*38), and so U satisfies the condition (*34). //

3.9 **Construction.** Given a direct system $(U_\alpha)_{\alpha \in \Lambda}$, $(\rho_{\alpha\beta})_{(\alpha,\beta) \in \Lambda_1}$ of sets, we can now construct a direct limit. Let

$$W = \amalg_{\alpha \in \Lambda} U_\alpha$$

be the disjoint union of all the sets U_α. On W define the relation $\sim$ by

$$u \sim v \iff \begin{cases} \text{if } u \in U_\alpha \text{ and } v \in U_\beta, \text{ then} \\ \exists \gamma \text{ with } \alpha \le \gamma,\ \beta \le \gamma \text{ and} \\ \text{such that } \rho_{\alpha\gamma}(u) = \rho_{\beta\gamma}(v). \end{cases} \qquad \begin{matrix} U_\alpha \searrow \\ \qquad U_\gamma \\ U_\beta \nearrow \end{matrix}$$

Then $\sim$ is an equivalence relation (clearly reflexive and symmetric; if $u \sim v$ and $v \sim w$ then we have

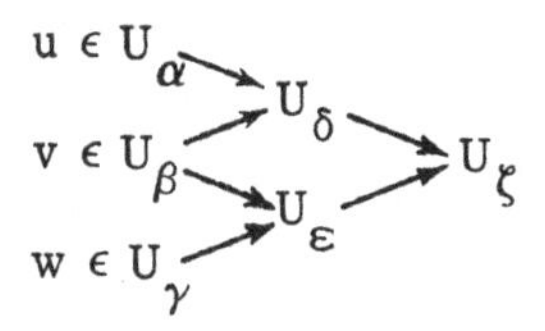

and we can pick ζ so that $\delta \le \zeta$, $\varepsilon \le \zeta$ to see that $u \sim w$).

Let $U = W/\sim$ and $\tau_\alpha : U_\alpha \to U$ be the composite maps $U_\alpha \hookrightarrow W \to W/\sim$.

3.10 Theorem. U with the $(\tau_\alpha)_{\alpha\epsilon\Lambda}$ is a direct limit for the system $(U_\alpha)_{\alpha\epsilon\Lambda}$. Hence every direct system of sets has a direct limit.

Proof. U, $(\tau_\alpha)_{\alpha\epsilon\Lambda}$ satisfy the conditions of Theorem 3.8. //

3.11 Definition. A direct system of abelian groups is a direct system of sets $(G_\alpha)_{\alpha\epsilon\Lambda}$, $(\rho_{\alpha\beta})_{(\alpha,\beta)\epsilon\Lambda_1}$ such that each G_α has an abelian group structure, and all the $\rho_{\alpha\beta}$ are homomorphisms with respect to these structures.

3.12 Example. As in Example 3.2 except we start with a presheaf of abelian groups.

3.14 Definition. A target for a direct system of abelian groups is a target G, $(\sigma_\alpha : G_\alpha \to G)_{\alpha\epsilon\Lambda}$ for the underlying direct system of sets, together with an abelian group structure on G such that all the σ_α are abelian group homomorphisms.

A direct limit is defined as in the set case, all maps involved being required to be abelian group homomorphisms.

3.15 Remark. Clearly we could define the same concepts relative to any structure e.g. direct limits of rings, groups, topological spaces, modules etc.

3.16 Proposition. Given a direct system of abelian groups, any two direct limits for it are naturally isomorphic (as abelian groups).

Proof. Exactly as in Proposition 3.6. //

3.18 Theorem. Suppose $(G_\alpha)_{\alpha\epsilon\Lambda}$, $(\rho_{\alpha\beta})_{(\alpha,\beta)\epsilon\Lambda_1}$ is a direct system of abelian groups and G, $(\tau_\alpha)_{\alpha\epsilon\Lambda}$ is a target satisfying

(i) $\forall g \in G \ \exists \alpha \in \Lambda$ such that $g \in \mathrm{Im}(\tau_\alpha)$

(ii) $\forall \alpha$, for $g_\alpha \in G_\alpha$ we have

$$\tau_\alpha(g_\alpha) = 0 \Longleftrightarrow \exists \beta \text{ such that } \alpha \leq \beta \text{ and } \rho_{\alpha\beta}(g_\alpha) = 0.$$

Then G is a direct limit for the system.

3.19 Construction. Given a direct system $(G_\alpha)_{\alpha\in\Lambda}$, $(\rho_{\alpha\beta})_{(\alpha,\beta)\in\Lambda_1}$ of abelian groups, let $H = \oplus_{\alpha\in\Lambda} G_\alpha$ be the direct sum of the G_α, with $i_\alpha : G_\alpha \to H$ the natural injections.

[Recall: H is the subgroup of $\Pi_{\alpha\in\Lambda} G_\alpha$ (with pointwise operations) generated by the images of all the i_α, where

$$(i_\alpha(g_\alpha))_\beta = \begin{cases} g_\alpha & \beta = \alpha \\ 0 & \beta \neq \alpha \end{cases} \qquad \text{for } g_\alpha \in G_\alpha .]$$

Let H_1 be the subgroup of H generated by all the

$$i_\alpha(g_\alpha) - i_\beta(\rho_{\alpha\beta}(g_\alpha))$$

as (α, β) runs through Λ_1 and (then) g_α runs through G_α. Let $G = H/H_1$ and $\tau_\alpha : G_\alpha \to G$ be the natural maps.

3.20 Theorem. G, $(\tau_\alpha : G_\alpha \to G)_{\alpha\in\Lambda}$ is a direct limit for the system. Hence any direct system of abelian groups has a direct limit.

Proofs of 3.18, 3.20 are very similar to those of 3.8 and 3.10, and are left as an exercise. //

3.21 Remark. In fact we could do without Construction 3.19 by constructing instead the set $\varinjlim$ and imposing an appropriate abelian group structure on it. But 3.19 works in more general circumstances (see 1. Ex. 9).

1.4 Stalks of presheaves

4.1 Let F be a presheaf (of sets or abelian groups) over a topological space X. Fix $x \in X$. The $F(U)$, as U runs through all open sets such that $U \ni x$, form a direct system with maps

$$\rho^U_V : F(U) \to F(V) \quad \text{whenever } U \supseteq V \ (\ni x).$$

Definition. The stalk F_x of F at x is $\varinjlim_{U\ni x} F(U)$. This comes equipped with maps

$$F(U) \to F_x : s \mapsto s_x$$

whenever an open $U \ni x$. The members of F_x are sometimes called germs (of sections of F).

4.2 **Proposition.** (a) Each germ $t \in F_x$ arises as $t = s_x$ for some $s \in F(U)$ for some open neighbourhood U of x.

(b) Two germs $s_x, t_x \in F_x$ (with $s \in F(U)$, $t \in F(V)$ say) are equal

$$s_x = t_x \Longleftrightarrow \exists \text{ open } W \subseteq U \cap V \text{ such that } \rho^U_W(s) = \rho^V_W(t).$$

Proof. This is just a restatement of Theorems 3.8 and 3.18, taking account of Proposition 3.6. //

4.3 **Examples.** A. For a constant presheaf A_X over X we have $A_{X,x} = A$ for each $x \in X$.

B. For a presheaf of functions such as C^Y, C^r or C^ω each germ at $x \in X$ extends to a function on some neighbourhood of x, and two germs are equal iff corresponding functions agree on some neighbourhood of x. Hence a germ summarises the 'local' behaviour of a function at a point.

C. For the pathological example P_1 of 2.E we have

$$\forall x \in X \qquad P_{1,x} = \{0\}$$

and yet it is not the constant presheaf $\{0\}_X$.

1.5 Morphisms of presheaves

5.1 **Definition.** If F. G are presheaves of sets over X, a morphism $f : F \to G$ is given by maps

$$f(U) : F(U) \to G(U)$$

for each open set U of X, such that whenever $U \supseteq V$ are open in X, the diagram

$$\begin{array}{ccc} F(U) & \xrightarrow{f(U)} & G(U) \\ {\scriptstyle \rho^U_V}\downarrow & & \downarrow{\scriptstyle \rho'^U_V} \\ F(V) & \xrightarrow{f(V)} & G(V) \end{array}$$

commutes, i. e.

$$\rho'^U_V f(U) = f(V)\rho^U_V.$$

If F, G are presheaves of abelian groups, for f to be a morphism of presheaves of abelian groups we require each $f(U)$ to be a homomorphism of abelian groups.

Composition of such morphisms is defined in the obvious way: $(g \circ f)(U) = g(U) \circ f(U)$ if $F \xrightarrow{f} G \xrightarrow{g} H$. As usual, we say that $f : F \to G$ is an isomorphism of presheaves (of sets or abelian groups) iff there is a morphism $g : G \to F$ such that $f \circ g = id_G$ and $g \circ f = id_F$ (where $id_F : F \to F$ is defined by $id_F(U) = id_{F(U)}$ for each open U in X).

5.2 **Proposition.** $f : F \to G$ is an isomorphism of presheaves (of sets or abelian groups)

iff $\forall$ open U of X $f(U)$ is an isomorphism
iff $\forall$ open U of X $f(U)$ is bijective.

Proof. f isomorphism $\Leftrightarrow \exists g$ such that $f \circ g = id_G$ and $g \circ f = id_F$
$\Leftrightarrow \exists g \, \forall U$ $f(U) \circ g(U) = id_{G(U)}$ and $g(U) \circ f(U) = id_{F(U)}$
$\Leftrightarrow \forall U$ $f(U)$ isomorphism.

For if $f : F \to G$ is a morphism with all $f(U)$ isomorphisms, the inverses $f(U)^{-1} : G(U) \to F(U)$ satisfy the conditions of compatibility with restriction. [This point needs checking.] //

5.3 **Remark.** We shall investigate later the extent to which this Proposition has meaning and is true with 'isomorphism' replaced by 'monomorphism' and 'epimorphism'.

5.4 **Construction.** Given a morphism of presheaves $f : F \to G$ on X, for each point $x \in X$ we can produce a morphism of stalks

$$f_x : F_x \to G_x$$

in such a way that whenever $F \xrightarrow{f} G \xrightarrow{g} H$ we have

$$(g \circ f)_x = g_x \circ f_x.$$

Given $x \in X$ we define f_x as follows: any $e \in F_x$ is of the form $e = s_x$ for some open $U \ni x$ and some $s \in F(U)$ (by 4.2); set $f_x(e) = (f(U)(s))_x$ (i.e. take the germ of the image of s). If also $e = s_x = t_x$ with $t \in F(V)$, then by 4.2 $\exists W \subseteq U \cap V$ with $x \in W$ and $\rho^U_W(s) = \rho^V_W(t)$; so

$$\rho^U_W(f(U)(s)) = f(W)\rho^U_W(s) = f(W)\rho^V_W(t) = \rho^V_W(f(V)(t))$$

so that $(f(U)(s))_x = (f(V)(t))_x$ and f_x is well-defined. The functoriality $(g \circ f)_x = g_x \circ f_x$ is easy to check; note that $(id_F)_x = id_{F_x}$ is obvious too.

5.5 **Remark.** 5.4 is in fact a special case of a generality concerning 'maps of direct systems'.

Exercises on Chapter 1

1. Prove that the $\varinjlim$ of the direct systems of the examples (3.2 and 3.12) is $F(\emptyset)$ in each case. Generalise.

2. Prove directly from the definitions (i.e. without Theorem 3.8 or the Construction 3.9) that if U is a direct limit of a direct system $(U_\alpha)_{\alpha \in \Lambda}$ of sets, then

$$U = \cup_{\alpha \in \Lambda} \text{Image}(U_\alpha \to U).$$

3. (i) Interpret and prove: a set is the direct limit of its finite subsets.

(ii) Interpret and prove: an abelian group is the direct limit of its finitely generated subgroups.

(iii) Can you obtain Z as a direct limit of finite abelian groups?

4. (i) Characterise direct systems of sets with $\varinjlim = \emptyset$.

(ii) Produce an interesting direct system of abelian groups with $\varinjlim = \{0\}$, the trivial group. Characterise such systems.

5. What can you say about the direct limit of a direct system all of whose maps are injective? Surjective?

6. For $n \in \mathbf{N}^*$, let $C_n(x)$ denote a cyclic group of order n with generator x. Let $p \in \mathbf{N}$ be a prime number. Let G be the direct limit of the following direct system of abelian groups:

$$\{0\} = C_{p^0}(x_0) \to C_p(x_1) \to C_{p^2}(x_2) \to \ldots \to C_{p^n}(x_n) \to C_{p^{n+1}}(x_{n+1}) \to \ldots$$

(where $C_{p^n}(x_n) \to C_{p^{n+1}}(x_{n+1})$ takes $x_n \mapsto px_{n+1}$). Preferably without resorting to the explicit construction prove:

(i) G is infinite, but torsion (i.e. every element has finite order).

(ii) Every finitely-generated subgroup of G is finite. Find all of them.

Deduce that G has no proper infinite subgroup, and no maximal proper subgroup. Can either of these situations arise for subspaces of a vector space (using dimension instead of order)? Identify a realisation of G inside the unit circle $\subseteq \mathbf{C}$ (under $\times$).

7. Consider the following direct system of abelian groups: fix $r \in \mathbf{Z}$; for all $n \in \mathbf{N}$ let $U_n = \mathbf{Z}$ and for $n \geq m$ let $\rho_{mn} : U_m \to U_n$ be multiplication by r^{n-m}. Identify the $\varinjlim$ as a subring of $\mathbf{Q}$.

8. Interpret and prove: the direct limit of a system of exact sequences is exact.

9. The notions of target and direct limit can be formulated without the restriction (a) of Definitions 3.1 and 3.11. What difference does this make to the Constructions? Find a system of abelian groups (in this generalised sense) with direct limit $A \oplus B$ without having this abelian group appear in the system. Justify Remark 3.21.

10. Formulate the dual notions of inverse system and inverse limit $\varprojlim$ (reverse the arrows). Find inverse systems:

(a) of finite sets whose $\varprojlim$ is infinite

(b) of finite abelian groups whose $\varprojlim$ is infinite

(c) of abelian groups whose $\varprojlim$ is $\mathbf{Z}$ (without $\mathbf{Z}$ in the system).

11. Verify that if (R_α) is a direct system of abelian groups such that each R_α is a ring and all the $\rho_{\alpha\beta}$ are ring morphisms, then $\varinjlim R_\alpha$ has a natural ring structure such that all the maps $R_\beta \to \varinjlim R_\alpha$ are ring morphisms.

12. What are the stalks of the presheaf P_2 of 2.E?

13. Construct a topological space X and presheaf F of abelian groups on X with the properties:

(a) for any open $U \subseteq X$ $F(U) \neq \{0\}$

(b) for all $x \in X$ the stalk $F_x = \{0\}$.

(If you cannot, prove that it is impossible.) (Compare Q4(ii).)

2 · Sheaves and sheaf spaces

We now study presheaves which satisfy additional axioms concerning the existence and uniqueness of sections with prescribed local nature (in the form of a given set of restrictions). In particular, we find that the geometric examples of presheaves of functions are in fact sheaves. We show that sheaves may be viewed as local homeomorphisms over the base space, and deduce that every presheaf gives rise to a sheaf in a universal way.

2.1 The sheaf axiom

1.1 **Definition.** Let X be a topological space and F a presheaf of sets over X. F is called a monopresheaf (or separated presheaf) iff it satisfies the condition (M):

(M) Suppose that U is an open set of X and $U = \cup_{\lambda\in\Lambda} U_\lambda$ is an open covering of U (i.e. each U_λ open in X), and $s, s' \in F(U)$ are two sections of F such that

$$\forall\lambda \in \Lambda \quad \rho^U_{U_\lambda}(s) = \rho^U_{U_\lambda}(s') ;$$

then $s = s'$.

1.2 **Examples.** The presheaves of 1.2.A-1.2.D are monopresheaves, but P_1 of 1.2.E is not.

1.3 We have also a 'glueing' condition (G):

(G) Suppose that U is open in X and $U = \cup_{\lambda\in\Lambda} U_\lambda$ is an open covering of U; suppose we are given a family $(s_\lambda)_{\lambda\in\Lambda}$ of sections of F with $\forall\lambda \in \Lambda$ $s_\lambda \in F(U_\lambda)$, such that

$$\forall\lambda, \mu \in \Lambda \quad \rho^{U_\lambda}_{U_\lambda\cap U_\mu}(s_\lambda) = \rho^{U_\mu}_{U_\lambda\cap U_\mu}(s_\mu);$$

then there is $s \in F(U)$ such that

$$\forall \lambda \in \Lambda \quad \rho^U_{U_\lambda}(s) = s_\lambda .$$

In other words, if the system (s_λ) is given on a covering and is consistent on all the overlaps, then it comes from a section over all of U.

1.4 Definition. A presheaf of sets over X satisfying (M) and (G) is called a sheaf of sets.

Similarly, an abelian sheaf is a presheaf of abelian groups which satisfies (M) and (G).

1.5 Remark. For a presheaf of abelian groups, we can simplify (M) by putting $s' = 0$.

1.6 There is a neat way to summarise the conditions (M) and (G). Given an open cover of an open set $U = \cup_{\lambda\in\Lambda} U_\lambda$ we can define maps

$$(*16) \quad F(U) \xrightarrow{a} \Pi_{\lambda\in\Lambda} F(U_\lambda) \underset{c}{\overset{b}{\rightrightarrows}} \Pi_{(\lambda,\mu)\in\Lambda\times\Lambda} F(U_\lambda \cap U_\mu)$$

by:

$$a(s) = (\rho^U_{U_\lambda}(s))_{\lambda\in\Lambda}$$

$$b((s_\lambda)_{\lambda\in\Lambda}) = (\rho^{U_\lambda}_{U_\lambda\cap U_\mu}(s_\lambda))_{(\lambda,\mu)\in\Lambda\times\Lambda}$$

$$c((s_\lambda)_{\lambda\in\Lambda}) = (\rho^{U_\mu}_{U_\lambda\cap U_\mu}(s_\mu))_{(\lambda,\mu)\in\Lambda\times\Lambda} .$$

If $A \xrightarrow{a} B \underset{c}{\overset{b}{\rightrightarrows}} C$ is a diagram of sets and maps, we say that a is an equaliser of (b, c) iff $\begin{cases} a \text{ is injective, and} \\ \text{Image}(a) = \{x \in B;\ b(x) = c(x)\} \end{cases}$ (i.e. iff A bijects with the subset of B on which b and c are equal).

Then we have:

1.7 Proposition. A presheaf F is a sheaf iff whenever $U = \cup_{\lambda\in\Lambda} U_\lambda$ is an open cover of an open set, the associated diagram (*16) of sets is an equaliser diagram (i.e. a is an equaliser of (b, c)).

Proof. An easy translation of the definition. //

1.8 Remark. If F is a presheaf of abelian groups, then the maps a, b, c are group morphisms and the equaliser condition is that the sequence of abelian groups

$$0 \to F(U) \xrightarrow{a} \Pi_{\lambda \in \Lambda} F(U_\lambda) \xrightarrow{b-c} \Pi_{(\lambda,\mu) \in \Lambda \times \Lambda} F(U_\lambda \cap U_\mu)$$

be _exact_ (that is, kernel = image at each point).

1.9 Exercise. Show that if G is an abelian sheaf, then

$G(\emptyset) = \{0\}$, the trivial group.

1.10 Proposition. _If F is a presheaf and G a monopresheaf over X, and $f, g : F \to G$ are two morphisms such that_

$\forall x \in X \quad f_x = g_x \quad$ (_i.e._ f, g _agree on all stalks_)

then $f = g$.

Proof. Let U be open in X and $s \in F(U)$. Then $f(s), g(s) \in G(U)$ ($f(s)$ is shorthand for $f(U)(s)$) and we wish to prove that $f(s) = g(s)$. Now

$$\forall x \in U \qquad f_x(s_x) = g_x(s_x)$$

that is

$$(f(s))_x = (g(s))_x$$

and so x has an open neighbourhood $U_x \subseteq U$ such that

$$\rho^U_{U_x}(f(s)) = \rho^U_{U_x}(g(s)).$$

Applying the condition (M) for G to the covering $(U_x)_{x \in U}$ of U we see that $f(s) = g(s)$. //

1.11 Remark. We shall mainly be interested in Proposition 1.10 when G is a sheaf (or even both F, G are sheaves).

1.12 Exercise. Find an example with G not a monopresheaf where 1.10 fails.

1.13 **Definition.** If F, G are sheaves (of sets or abelian groups) and $f : F \to G$ is a presheaf morphism, we also call f a morphism of sheaves. A moment's thought and Proposition 1.5.2 show that if $F \to G$ is an isomorphism of presheaves, and G is a sheaf, then F is a sheaf.

2.2 Examples of sheaves

2.A The presheaf P_2 of 1.2.E is a sheaf.

2.B All the examples of presheaves of functions C^Y, C^r, C^ω of 1.2.B-D are in fact sheaves, and all for the same reason: in order for an arbitrary map to Y, $\mathbf{R}$, $\mathbf{C}$ to satisfy the appropriate condition (continuity, differentiability, analyticity) it is necessary and sufficient that the condition be satisfied in some small neighbourhood of each point: thus a 'glued' function will also satisfy the condition.

2.C Let E be a topological space and $p : E \to X$ a continuous map. We can construct a sheaf F of sections of p: for U open in X let

$$F(U) = \{\text{continuous maps } \sigma : U \to E;\ \begin{array}{ccc} & \overset{\sigma}{\nearrow} & E \\ & & \downarrow p \\ U & \hookrightarrow & X \end{array} \text{ commutes, i.e. } p \circ \sigma = \mathrm{id}_U\}$$

If $U \supseteq V$ are open in X we have an easy restriction:

$$\rho^U_V : F(U) \to F(V) : \sigma \mapsto \sigma | V.$$

The conditions (M) and (G) are easy to verify.

Note that the condition on σ means that σ gives for each $x \in U$ a (continuous) choice of a point $\sigma(x) \in p^{-1}(x)$, the fibre of p over x.

2.3 Sheaf spaces

In 2.3 and 2.4 we shall be concerned with presheaves of sets.

3.1 **Proposition.** If X is a topological space and F a sheaf on X, then for any open U and $s, s' \in F(U)$ we have

$$s = s' \iff \forall x \in U \quad s_x = s'_x \,;$$

that is, two sections of F agree iff their germs are everywhere equal.

Proof. $\Rightarrow$ is clear. Conversely, given $s, s' \in F(U)$ such that $\forall x \in U \ s_x = s'_x$, for each $x \in U$ we can find an open $U_x \ni x$ such that $\rho^U_{U_x}(s) = \rho^U_{U_x}(s')$. Applying the condition (M) to $(U_x)_{x \in U}$ we see that $s = s'$. //

3.2 **Remark.** This is not true for arbitrary presheaves, e.g. the presheaf P_1 of 1.2.E (cf. 1.4.3). But we are led to try to represent a sheaf as a collection of functions with values in its stalks.

3.3 **Definition.** Let X be a topological space. A sheaf space over X is a pair (E, p) of a topological space E and a continuous map $p : E \to X$ such that p is a local homeomorphism, that is:
$\forall y \in E \ \exists$ open $N \ni y$, open $U \ni p(y)$ such that $p|N : N \to U$ is a homeomorphism.

A morphism of sheaf spaces $f : (E, p) \to (E', p')$ is a continuous map $f: E \to E'$ such that

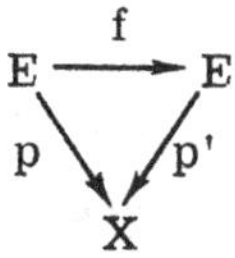

$$\begin{array}{ccc} E & \xrightarrow{f} & E' \\ & {\scriptstyle p}\searrow \quad \swarrow{\scriptstyle p'} & \\ & X & \end{array}$$

commutes i.e. $p = p' \circ f$.

3.4 **Construction.** For each sheaf space E we can construct a sheaf of sets ΓE (the sheaf of sections of E) in such a way that a morphism $f : E \to E'$ of sheaf spaces gives rise to a morphism $\Gamma f : \Gamma E \to \Gamma E'$ of sheaves.

We have seen in 2.2.C how to construct the sheaf of sections of the pair (E, p). We let, for U open in X,

$$\Gamma(U, E) = \{\text{continuous maps } \sigma : U \to E;\ p \circ \sigma = id_U\} \qquad \begin{array}{ccc} & & E \\ & \sigma \nearrow & \downarrow p \\ U & \hookrightarrow & X \end{array}$$

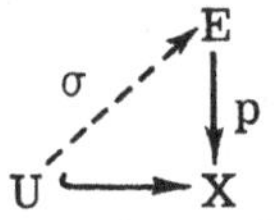

and then the presheaf $\Gamma E : U \mapsto \Gamma(U, E)$ is a sheaf.

Given a morphism $f : E \to E'$ of sheaf spaces, we obtain

$$\begin{array}{l} \Gamma(U, E) \to \Gamma(U, E') \\ \quad \sigma \mapsto f \circ \sigma \end{array} \qquad \begin{array}{ccc} & E \xrightarrow{f} E' & \\ \sigma \nearrow & {\scriptstyle p}\searrow \ \swarrow{\scriptstyle p'} & \\ U & \hookrightarrow X & \end{array}$$

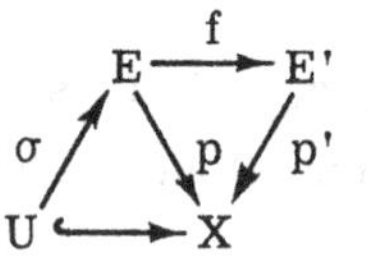

and this gives a morphism of sheaves $\Gamma f : \Gamma E \to \Gamma E'$.

3.5 Lemma. *Let* (E, p) *be a sheaf space over* X. *Then*

(a) p *is an open map*

(b) *if* U *is open in* X *and* $\sigma \in \Gamma(U, E)$, *then* $\sigma[U]$ *is open in* E; *furthermore sets of this form give a basis for the topology of* E.

(c) *If* $E \xrightarrow{\phi} E'$, with $p: E \to X$ and $p': E' \to X$, *is a commutative diagram of maps, and* p, p' *are local homeomorphisms, then* ϕ *continuous* $\Leftrightarrow$ ϕ *open* $\Leftrightarrow$ ϕ *local homeomorphism.*

Proof. (a) Let W be open in E and $x \in p[W]$. Pick any $e \in W$ such that $p(e) = x$. Then by the definition of a sheaf space, e has an open neighbourhood $W' \subseteq W$ mapped by p onto an open set in X; i.e. x has an open neighbourhood $p[W']$ inside $p[W]$.

(b) Any $e \in \sigma[U]$ has an open neighbourhood $W \subseteq E$ such that $p|W$ is a homeomorphism onto an open set $V \subseteq X$. Then $p|W$ maps $W \cap \sigma[U]$ bijectively to $U \cap V$, which is open in X; hence $W \cap \sigma[U]$ is an open neighbourhood of e inside $\sigma[U]$. The last part is easy, using (a).

(c) By definition and part (a), local homeomorphism $\Rightarrow$ continuous and open. Let us prove ϕ continuous $\Rightarrow$ ϕ local homeomorphism. Given $y \in E$, $\phi(y) \in E'$ and so since p' is a local homeomorphism there are open N', V such that $\phi(y) \in N' \xrightarrow{p'|N'} V$ is a homeomorphism. Also $\phi^{-1}(N')$ is open in E, so we can construct a diagram

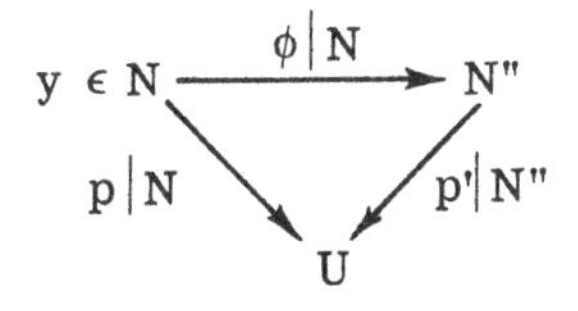

with N, N'', U open and $p|N$, $p'|N''$ both homeomorphisms. Hence $\phi|N$ is a homeomorphism.

ϕ open $\Rightarrow$ ϕ local homeomorphism is similar (first pick a suitable neighbourhood of y). //

3.6 Proposition. *If* (E, p) *is a sheaf space, then the stalk of* ΓE *at* $x \in X$ *is (up to natural bijection) just the fibre* $p^{-1}(x)$ *of* p *over* x, <u>which has the discrete topology as a subspace of</u> E.

Proof. For $x \in U$ open in X we have the map

(*36) $\Gamma E(U) = \Gamma(U, E) \to p^{-1}(x) : \sigma \mapsto \sigma(x)$

(since σ maps $U \to E$); these maps are clearly compatible with restrictions. Thus we wish to prove that this target is a direct limit, and we use the criterion of 1. 3. 8:

(a) each $e \in p^{-1}(x)$ arises as an image under one of these maps; for, since p is a local homeomorphism e has a neighbourhood W in E such that $p|W : W \to U$ is a homeomorphism, with U open in X. The inverse $\sigma = (p|W)^{-1}$ of this map has $\sigma \in \Gamma(U, E)$ and maps to e under (*36).

(b) if $s \in \Gamma(U, E)$ and $t \in \Gamma(V, E)$ agree at x, by Lemma 3. 5 $W = s[U] \cap t[V]$ is open in E, and s, t agree on $p[W]$ (which is open by Lemma 3. 5) since they are both inverses of $p|W$. Hence

$$\rho^U_{p[W]}(s) = \rho^V_{p[W]}(t) \in \Gamma(p[W], E).$$

Thus

$$p^{-1}(x) \cong \varinjlim_{U \ni x} \Gamma(U, E).$$

To see that $p^{-1}(x)$ is a discrete subspace of E, note that for $e \in p^{-1}(x)$ and W as constructed in (a) above, W is open and $W \cap p^{-1}(x) = \{e\}$. //

3. 7 **Exercise.** Check the functorial properties $\begin{cases} \Gamma(f \circ g) = \Gamma f \circ \Gamma g \\ \Gamma(id) = id. \end{cases}$

Check also that if $f : E \to E'$ is a morphism of sheaf spaces over X, then

$$(\Gamma f)_x : (\Gamma E)_x \to (\Gamma E')_x$$

and

$$f|p^{-1}(x) : p^{-1}(x) \to p'^{-1}(x)$$

are isomorphic maps.

3.8 Construction. For each presheaf F on X we can construct a sheaf space LF in such a way that any morphism $f : F \to F'$ of presheaves gives rise to a morphism $Lf : LF \to LF'$ of sheaf spaces.

Set $LF = \amalg_{x \in X} F_x$ (the disjoint union of the stalks of F) with $p : LF \to X$ the natural projection, so that $p^{-1}(x) = F_x$. We topologise LF as follows: let U be open in X and $s \in F(U)$; then we can define a map

$$\hat{s} : U \to LF : x \mapsto s_x \in F_x .$$

We prescribe that all the sets $\hat{s}[U] = \{s_x \in LF;\ x \in U\}$ be open sets. Then $\{\hat{s}[U];\ s \in F(U)\}$ forms a basis for the topology it generates, for

$$e \in \hat{s}[U] \cap \hat{t}[V] \quad (\text{with } s \in F(U),\ t \in F(V))$$

$\Rightarrow$ s, t agree in germ at $p(e) = x$ say

$\Rightarrow$ s, t agree in a neighbourhood W of x (with $W \subseteq U \cap V$)

$\Rightarrow$ e has a basic neighbourhood $\hat{s}[W] = \hat{t}[W]$ inside $\hat{s}[U] \cap \hat{t}[V]$ (where we really mean $\widehat{\rho_W^U(s)}[W]$ by $\hat{s}[W]$).

Furthermore p is continuous with respect to this topology on LF, since for any open U of X

$$p^{-1}(U) = \cup\{\hat{s}[V];\ s \in F(V) \text{ with } V \subseteq U \text{ open}\},$$

and p is a local homeomorphism since on $\hat{s}[U]$ it has the continuous inverse $\hat{s}$.

A presheaf morphism $f : F \to F'$ gives a collection of stalk maps $f_x : F_x \to F'_x$ and so a map $Lf : LF \to LF'$ such that

$$\begin{array}{ccc} LF & \longrightarrow & LF' \\ & {\scriptstyle p}\searrow \quad \swarrow{\scriptstyle p'} & \\ & X & \end{array}$$

commutes; also $Lf[\hat{s}[U]] = \widehat{f(U)(s)}[U]$, so Lf is continuous by 3.5(c).

3.9 Exercise. Check the functorial properties $\begin{cases} L(f \circ g) = Lf \circ Lg \\ L(id) = id. \end{cases}$

Now it is natural to ask what happens when we do L, Γ in succession.

3.10 Theorem. <u>If</u> E <u>is a sheaf space over</u> X, <u>then</u> $L\Gamma E$ <u>is isomorphic to</u> E <u>as sheaf spaces over</u> X <u>(i.e. there is a morphism</u>

$\phi : E \to L\Gamma E$ *of sheaf spaces with a two-sided inverse).*

Proof. Fix nomenclature:

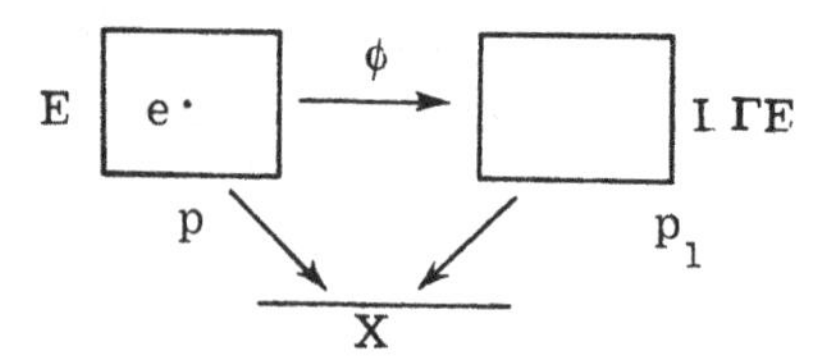

For $x \in X$, the fibre $p^{-1}(x)$ bijects with the stalk of ΓE at x, and so with the fibre $p_1^{-1}(x)$ in $L\Gamma E$. These bijections fit together to give a bijection ϕ such that $p = p_1 \circ \phi$.

If U is open in X and $\sigma \in \Gamma(U, E)$, then

$$\phi[\sigma[U]] = \hat{\sigma}[U].$$

Hence ϕ is open, and by 3.5 it is also continuous; since ϕ is bijective this means that ϕ is a homeomorphism. //

2.4 The sheafification of a presheaf

4.1 Given a presheaf F over X we can construct the sheaf space LF and then obtain a *sheaf* ΓLF called the *sheafification* of F. Now we have a morphism of presheaves $n_F : F \to \Gamma LF$ defined as follows: given U open in X and $s \in F(U)$, s defines the function

$$\hat{s} : U \to LF : x \mapsto s_x \in F_x$$

as in 3.8 and so $\hat{s} \in \Gamma(U, LF)$; then $n_F(U) : s \mapsto \hat{s}$.

Now this construction enjoys the following universal property:

4.2 **Theorem.** *Let* F *be a presheaf and* G *a sheaf over* X. *Then any morphism of presheaves* $f : F \to G$ *factors uniquely through* $F \to \Gamma LF$; *that is, given* $F \xrightarrow{f} G$, $F \searrow \Gamma LF$, *there is a unique sheaf morphism* $\Gamma LF \xrightarrow{g} G$ *making*

$$\begin{array}{ccc} F & \xrightarrow{f} & G \\ & {}_{n_F}\searrow \quad \nearrow_{g} & \\ & \Gamma LF & \end{array}$$

commute.

This theorem gives precise meaning to the notion that $\Gamma L F$ is the 'best' sheaf that can be made out of F.

Before the proof we need two lemmas.

4.3 Lemma. If (and only if) G is a sheaf, then (and only then) $G \to \Gamma L G$ is an isomorphism of sheaves.

Proof. 'Only if' is clear, by 1.13. For 'if' we check that each $G(U) \to \Gamma L G(U) = \Gamma(U, LG) : s \mapsto \hat{s}$ (for U open in X) is bijective; this is enough by Proposition 1.5.2.

(i) The map is injective; for by Proposition 3.1

$$\hat{s} = \hat{s}' \iff \forall x \in U \quad s_x = s'_x \iff s = s'.$$

(ii) The map is surjective; for given $t \in \Gamma(U, LG)$, $t[U]$ is open in LG by Lemma 3.5. For each $x \in U$, $t(x) \in G_x$ has a basic neighbourhood inside $t[U]$, of the form $\hat{s}^x[U_x]$ for some open $U_x \subseteq U$ and some $s^x \in G(U_x)$ (by 3.8). This means that the $s^x \in G(U_x)$ satisfy the glueing condition (G) of 1.3; for in $V = U_x \cap U_y$, $\rho_V^{U_x}(s^x)$ and $\rho_V^{U_y}(s^y)$ have the same germ everywhere (viz. $t(z)$ at $z \in V$) and so are equal by Proposition 3.1. Hence since G is a sheaf $\exists s \in G(U)$ such that $\forall x \in U \;\; s_x = (s^x)_x = t(x)$ i.e. $\hat{s} = t$. //

4.4 Remark. The morphism $n_F : F \to \Gamma L F$ (and so the isomorphism of 4.3 when F is a sheaf) is natural in the sense that if $f : F \to F'$ is a morphism of presheaves, then the following diagram commutes:

$$\begin{array}{ccc} F & \xrightarrow{n_F} & \Gamma L F \\ {\scriptstyle f}\downarrow & & \downarrow{\scriptstyle \Gamma L f} \\ F' & \xrightarrow[n_{F'}]{} & \Gamma L F' \end{array}$$

4.5 Lemma. For any presheaf F on X, all the maps

$$F_x \to (\Gamma L F)_x$$

induced on stalks by n_F are isomorphisms.

Proof. Clear, since the stalk $(\Gamma LF)_x$ is naturally the fibre of LF over x, which is F_x (by 3.6 and 3.8), and the map $F_x \to (\Gamma LF)_x$ is induced by the maps (*36) of Proposition 3.6 . //

Proof of Theorem 4.2. If $g : \Gamma LF \to G$ exists making

$$\begin{array}{ccc} & & G \\ & \overset{f}{\nearrow} & \uparrow g \\ F & \underset{n_F}{\searrow} & \Gamma LF \end{array}$$

commute, then its stalk maps g_x for $x \in X$ are determined as the composites

$$(\Gamma LF)_x \underset{n^{-1}_{F,x}}{\rightleftarrows} F_x \xrightarrow{f_x} G_x \qquad (n^{-1}_{F,x} \text{ exists by Lemma 4.5});$$

hence g is determined uniquely, by 1.10.

To show that g exists, we take $F \xrightarrow{f} G$ and apply 3.8 to get $LF \xrightarrow{Lf} LG$, and then apply 3.4 to get

$$\Gamma LF \xrightarrow{\Gamma Lf} \Gamma LG \xrightarrow{\sim} G,$$

by Lemma 4.3. It is easy to check that the resulting triangle commutes (use 4.4). //

4.6 If A is a given set, what should we call the constant sheaf A over X? If we take the constant presheaf A_X of 1.2.A and apply the sheafification procedure ΓL, we first obtain a sheaf space $LA_X \xrightarrow{p} X$ such that

$$\forall x \in X \qquad p^{-1}(x) = A_{X,x} = A$$

by 1.4.3A so that set-theoretically $LA_X = A \times X$ and $p = \pi_2$ (projection on second factor). By 3.8 the topology on $A \times X$ has as basis sets of the form $\{a\} \times U$ for $a \in A$ and U open in X; hence it is the product topology, with A given the discrete topology.

The sections of $F = \Gamma LA_X$ are given by

$F(U) = \Gamma(U, LA_X)$ = set of continuous functions σ making

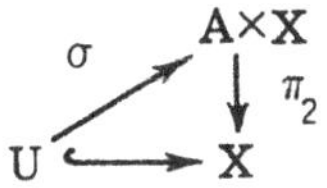

$$\begin{array}{ccc} & & A\times X \\ & \overset{\sigma}{\nearrow} & \downarrow \pi_2 \\ U & \hookrightarrow & X \end{array}$$

commute

$\cong$ set of continuous functions $s : U \to A$ (with A discrete)

$\cong$ set of locally constant functions $s : U \to A$.

Note that for U disconnected (and A with > 1 element) this has $F(U) \neq A$, so that A_X was not a sheaf originally by 4.3.

Definition. The constant sheaf over X modelled on A is the sheaf whose sheaf space is $A \times X \xrightarrow{\pi_2} X$ (A given discrete topology). It is sometimes denoted (confusingly) by A_X.

2.5 Sheaf spaces of abelian groups

5.1 We now wish to define a sheaf space of abelian groups and constructions Γ, L in such a way as to make Theorem 4.2 (in particular) true for sheaves and presheaves of abelian groups.

Given a presheaf of abelian groups F over X, each stalk is an abelian group and so the corresponding sheaf space (LF, p) has the property:

(a) For each $x \in X$, the fibre $p^{-1}(x)$ is an abelian group.

However, the abelian group structures also 'vary continuously' as x varies; more precisely:

5.2 **Proposition.** For a sheaf space (E, p) satisfying (a) of 5.1 the following two conditions are equivalent:

(b) For any U open in X the set $\Gamma(U, E)$ is an abelian group under pointwise addition of functions;

(b') Let $E\pi E = \{(e, e') \in E \times E;\ p(e) = p(e')\}$; then the map $m : E\pi E \to E : (e, e') \mapsto e - e'$ is continuous (where $-$ denotes subtraction in $p^{-1}(p(e))$).

Proof. (b') $\Rightarrow$ (b): We have the diagram

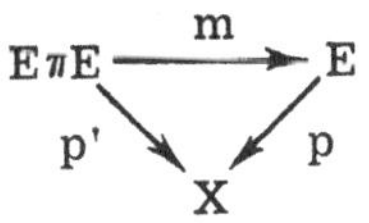

$$\begin{array}{ccc} E\pi E & \xrightarrow{m} & E \\ & {\scriptstyle p'}\searrow \quad \swarrow{\scriptstyle p} & \\ & X & \end{array}$$

where $p' : (e, e') \mapsto p(e) = p(e')$. If $f, g \in \Gamma(U, E)$ then $f-g : U \to E$ can be written as the composite

$$\begin{array}{ccccc} & (f, g) & & m & \\ U & \to & E\pi E & \to & E \\ x & \mapsto & (f(x), g(x)) & & \end{array}$$

and so is continuous i.e. $f - g \in \Gamma(U, E)$. The result follows easily.

(b) $\Rightarrow$ (b') By Lemma 3.5(b) we need to check that for any open $U \subseteq X$ and $f \in \Gamma(U, E)$ we have

$$m^{-1}(f[U]) \text{ open in } E\pi E.$$

But $(e, e') \in m^{-1}(f[U]) \Rightarrow$ for $x = p(e) = p(e')$, $e - e' = f(x)$. Pick by Proposition 3.6 an open W in X and $g, g' \in \Gamma(W, E)$ such that $g(x) = e$, $g'(x) = e'$. Then $g - g'$ and f agree in germ at x and so for some open $V \subseteq W \cap U$

$$\rho^W_V(g - g') = \rho^U_V(f)$$

by Proposition 3.6 again. Then (e, e') has the basic neighbourhood

$$(g[V] \times g'[V]) \cap E\pi E \quad \text{inside } m^{-1}(f[U]). \ /\!/$$

(Aliter use 3.5(c).)

5.3 **Remark.** (b') is often summarised by saying that subtraction is continuous on E.

5.4 **Definition.** A <u>sheaf space of abelian groups</u> over X is a sheaf space (E, p) satisfying condition (a) of 5.1 and conditions (b) and (b') of 5.2. A <u>morphism</u> $(E, p) \to (E', p')$ of such is a sheaf space morphism such that $\forall x \in X$ the map $p^{-1}(x) \to p'^{-1}(x)$ is a homomorphism of abelian groups.

5.5 With this definition and the results of 5.2 we see that the constructions Γ, L of 3.4 and 3.8 take us from sheaf spaces of abelian groups to abelian sheaves, and from presheaves of abelian groups to sheaf spaces of abelian groups.

The following results remain true in the context of sheaves, presheaves and sheaf spaces of abelian groups, and their appropriate morphisms:

3.5, 3.6, 3.7, 3.9, 3.10, 4.2, 4.3, 4.4, 4.5.

5.6 **Corollary.** _If_ F, G _are abelian sheaves over_ X, _and_ $f : F \to G$ _is a morphism as sheaves of sets, then_ f _is a morphism of abelian sheaves iff_ $\forall x \in X$ $f_x : F_x \to G_x$ _is a homomorphism of abelian groups._

Proof. 'Only if' is clear. If f has the above property, then $Lf : LF \to LG$ is a morphism of sheaf spaces of abelian groups; hence $\Gamma Lf : \Gamma LF \to \Gamma LG$ is a morphism of abelian sheaves; but by 4.3, 4.4 and 5.5 we see that this implies that f is too. //

5.7 **Terminology.** Some authors define a sheaf over a topological space to be what we have called a sheaf space over X. The existence of constructions Γ, L with properties 4.3 and 4.4 shows that there is no essential difference between these definitions (a sheaf determines a unique sheaf space, and conversely; and similarly for morphisms).

This leads to the following widely used notation: if F is a sheaf (of sets or abelian groups) over X and U is open in X, _we shall from now on_ write Γ(U, F) instead of F(U) for the set (or abelian group) of sections of F over U. By 4.3 $\Gamma(U, F) \cong \Gamma(U, LF)$, and this is the origin of the terminology 'sections of F over U'.

This approach involves thinking of a sheaf F as the collection of its stalks: hence some authors use the terminology 'sheaf of _germs_ of continuous (or differentiable, or analytic) functions' for the sheaves described in 1.2.B-D and 2.2B.

Exercises on Chapter 2

1. Let $I = [0, 1] \hookrightarrow \mathbf{R}$. Show that there is a unique (up to isomorphism) sheaf F on I with stalks:

$$F_0 = F_1 = \mathbf{Z}$$
$$F_x = \{0\} \text{ if } x \in I \setminus \{0, 1\}.$$

What is $\Gamma(I, F)$?

Let G be the constant sheaf $\mathbf{Z}$ on I (4.6). How many morphisms are there from F to G? From G to F?

2. Show that the following conditions are equivalent for a topological space X:

(a) X is locally connected (that is, each point has a base of connected neighbourhoods);

(b) for any set A, the constant sheaf A_X (4.6) has

$$\Gamma(U, A_X) = \Pi_{t \in U'} A$$

for U open in A where U' is the set of connected components of U;

(c) (b) holds for $A = \{0, 1\}$, some set with two elements.

When these conditions hold, what are the restriction maps in terms of the representation given in (b)?

[Hint, if necessary: Bourbaki, Gen. Top. Ch. I, §11.6, Prop. 11.]

3. Let F be a presheaf on a space X, and let V be open in X. Then we can define a presheaf $F|V$ on V by the same recipe as F; that is

$$(F|V)(U) = F(U)$$

for U open in V. Show that if F is a sheaf, so is $F|V$. Show also that if F has sheaf space $LF \xrightarrow{p} X$, then $F|V$ has sheaf space $(p^{-1}V, p|p^{-1}V)$. What can you say when V is not open?

(Compare Q4 below and §3.8.)

4. Let F be a sheaf on a space X with sheaf space $LF \xrightarrow{p} X$, and let A be a subspace of X. We can define the set (or abelian group) of sections of F over A by

$$\Gamma(A, F) = \Gamma(A, LF) = \text{set of sections of the continuous map}$$
$$p^{-1}A \xrightarrow{p} A.$$

(Compare 2.C and 3.4.) Show that we can define $\Gamma(A, F)$ in terms of F alone as

$$\Gamma(A, F) = \varinjlim \Gamma(U, F)$$

where the direct limit is taken over the set of open subsets U of X such that $U \supseteq A$. (Colloquially, this says that a section of F over A extends uniquely into a small neighbourhood of A.)

5. Let F be a sheaf on a space X and let $(M_i)_{i \in I}$ be a locally finite covering of X by closed sets (so that for each $x \in X$, $\{i \in I;\ x \in M_i\}$ is finite). In the notation of Q4, suppose we are given a family $(s_i)_{i \in I}$ with

$$\forall i \in I \quad s_i \in \Gamma(M_i, F)$$

and

$$\forall i, j \in I \ s_i = s_j \text{ on } M_i \cap M_j.$$

Show that there is a unique $s \in \Gamma(X, F)$ with

$$\forall i \in I \quad s = s_i \text{ on } M_i.$$

6. Let K be any infinite field and $L = K(t)$ a simple transcendental extension (= the field of fractions of the polynomial ring $K[t]$). Let X be the topological space obtained by giving K the topology whose closed sets are the finite subsets of K.

Define a sheaf $\mathcal{O}$ of commutative-rings-with-a-one on X as follows: for U open in X, $U \neq \emptyset$, let

$$(*) \qquad \mathcal{O}(U) = \{f \in L;\ \exists g, h \in K[t] \text{ with } f = \tfrac{g}{h} \text{ and } \forall P \in U\ h(P) \neq 0\} \subseteq L.$$

If $\emptyset \neq V \subseteq U$ then $\mathcal{O}(U) \subseteq \mathcal{O}(V) \subseteq L$ and we take the inclusion as the restriction map ρ^U_V. Show that $\mathcal{O}$ is a sheaf of rings on X (it is called the <u>sheaf of rational functions on the affine line</u> X <u>over</u> K). [Hint: first prove that X is compact.]

Identify the stalk $\mathcal{O}_P$ of $\mathcal{O}$ at $P \in X$ as a subring of L, and show that it is a <u>local</u> ring (i.e. has a unique maximal ideal m_P); what is its residue field $(= \mathcal{O}_P/m_P)$ and its field of fractions?

Show that the set of all non-empty open sets of X is directed by $\supseteq$ and that $\varinjlim \Gamma(U, \mathcal{O}) = L$.

When does $\mathcal{O}$ have non-polynomial global sections? That is, we certainly have $K[t] \subseteq \Gamma(X, \mathcal{O})$; when is the inequality strict? (Give a necessary and sufficient condition on K.)

We can consider $f \in \Gamma(U, \mathcal{O})$ as a function on U; namely, express $f = g/h$ as in (*) and define for $P \in U$

$$f(P) = g(P)/h(P) \in K.$$

Show that this defines a morphism $\phi : \mathcal{O} \to F$ where F is the sheaf of K-valued functions on X (giving K the indiscrete topology), and that putting $\mathcal{O}'(U) = \text{Image}(\phi(U))$ defines a sheaf $\mathcal{O}'$ with a morphism $\mathcal{O} \to \mathcal{O}'$. Prove that $\mathcal{O} \to \mathcal{O}'$ is an isomorphism of sheaves. Hence we may regard $\mathcal{O}$ as a sheaf of K-valued functions on X.

For $K = \mathbf{C}$ show that $\mathcal{O}'$ is a subsheaf of the sheaf C^{ω} of analytic C-valued functions on $X = \mathbf{C}$ (in a suitable sense).

[We shall see later (§4.2) that Q6 is a special case of a very powerful construction (the prime spectrum of a commutative ring) which will yield analogous results for $X = K^n$ instead of K, or even any subset X of K^n defined by polynomial equations.]

3 · Morphisms of sheaves and presheaves

In this chapter we first give an account of the elementary language of category theory, and show how this gives a unified way of looking at many of the ideas we have been considering. We are led to look for convenient properties of the categories of sheaves and of presheaves over a given topological space, and we find that they each have a list of such properties which are summarised in the definition of abelian category.

However, the construction of cokernels differs in the two categories; this expresses what is perhaps the basic question in sheaf theory: to what extent does a sheaf epimorphism (a map of sheaves which is 'locally' surjective) have surjective section maps? This is studied further when we consider cohomology (Chapter 5).

Lastly, we consider what happens in a change of base space by a continuous map. We find that there is a covariant (that is, going in the same direction as the map) method of changing the base space of presheaves, and a contraviant (opposite direction) construction which is a generalisation of sheafification. These are connected by an adjointness relation, which may be interpreted as expressing their universal nature. In the case of an inclusion map of a locally closed subspace, we also consider the process of extension by zero.

3.1 Categories and functors

1.1 Definition. A <u>category</u> C consists of

(a) a class ObC of <u>objects</u>

(b) for each A, B $\in$ ObC a set $\mathrm{Hom}_C(A, B)$ of <u>morphisms</u> from A to B

(c) for each A, B, D $\in$ ObC a function (<u>composition</u>)

$$\mathrm{Hom}_C(B, D) \times \mathrm{Hom}_C(A, B) \to \mathrm{Hom}_C(A, D)$$

written

$$(g, f) \mapsto g \circ f$$

such that

(i) for each $A \in \mathrm{Ob}C$ $\exists$ an <u>identity</u> $1_A \in \mathrm{Hom}_C(A, A)$ such that

$\forall B \in \mathrm{Ob}C$ $\forall f \in \mathrm{Hom}(A, B)$ $f \circ 1_A = f$

and $\forall f \in \mathrm{Hom}(B, A)$ $1_A \circ f = f$

(ii) whenever $A, B, D, E \in \mathrm{Ob}C$ and $\left.\begin{array}{l} f \in \mathrm{Hom}(A, B) \\ g \in \mathrm{Hom}(B, D) \\ h \in \mathrm{Hom}(D, E) \end{array}\right\}$ then $h \circ (g \circ f) = (h \circ g) \circ f$.

(<u>associativity</u>).

1.2 **Notation.** When $f \in \mathrm{Hom}_C(A, B)$ we write $f : A \to B$ or $A \xrightarrow{f} B$ and use diagrams in an obvious way; for instance, the hypothesis of (ii) is 'whenever $A \xrightarrow{f} B \xrightarrow{g} D \xrightarrow{h} E$'. We sometimes write $\mathrm{Mor}\, C = \coprod_{A, B \in \mathrm{Ob}C} \mathrm{Hom}_C(A, B)$ (disjoint union).

1.3 **Exercise.** Show that identities are unique.

1.4 **Examples and definitions.**

A. Letting $\mathrm{Ob}C$ = class of all $\left\{\begin{array}{l} \text{sets} \\ \text{abelian groups} \\ \text{topological spaces} \end{array}\right.$

and $\mathrm{Hom}_C(A, B)$ = set of all $\left\{\begin{array}{l} \text{maps} \\ \text{homomorphisms} \\ \text{continuous maps} \end{array}\right.$ and composition

to be composition of maps we obtain the categories Sets, Abgp, Top.

Similarly for any other type of mathematical structure e.g. category of groups; for a ring R the category of R-modules.

B. Given a preordered set Λ (i.e. a relation $\leq$ on Λ which is reflexive and transitive) we can consider Λ as a category C with

$$\mathrm{Ob}C = \Lambda$$
$$\mathrm{Hom}_C(\lambda, \mu) = \begin{cases} \text{singleton} & \text{if } \lambda \leq \mu \\ \emptyset & \text{if } \lambda \not\leq \mu \end{cases}$$

(composition is then uniquely determined).

C. For $X \in Ob(Top)$ we have defined three categories:
Presh (or Presh/X if we wish to emphasise X) has objects the presheaves of abelian groups over X, and morphisms the presheaf morphisms.
Shv (or Shv/X) has objects the abelian sheaves over X.
Shfsp (or Shfsp/X) has objects the sheaf spaces of abelian groups over X.
In each case the composition of morphisms is that with which we are familiar.
(Note: from now on we shall mainly be concerned with presheaves and sheaves of abelian groups, so we do not reserve special names for the categories of sheaves and presheaves of sets over X.)

1.5 **Definition.** Given two categories C, D a covariant functor $F : C \to D$ is given by:

(a) a map $F : ObC \to ObD$

(b) $\forall A, B \in ObC$ a map $F : Hom_C(A, B) \to Hom_D(FA, FB)$

such that

(i) $\forall A \in ObC$ $F(1_A) = 1_{FA}$

(ii) $\forall f, g \in MorC$ $F(f \circ g) = Ff \circ Fg$ whenever $f \circ g$ is defined (so RHS is too).

A contravariant functor has instead of (b)

(b') $\forall A, B \in ObC$ a map $F : Hom_C(A, B) \to Hom_D(FB, FA)$

and (ii) is replaced by

(ii') $\forall f, g \in MorC$ $F(f \circ g) = Fg \circ Ff$ whenever $f \circ g$ is defined (so RHS is too).

Functors can be composed in an obvious manner.

1.6 **Examples.**

A. There are functors Abgp $\to$ Sets, Top $\to$ Sets sending an object to its underlying set (called forgetful functors).

B. There is an inclusion functor Shv/X $\to$ Presh/X.

C. If Λ is a directed set (1.3.1) and we consider Λ as a category as in 1.4 B, then a direct system of sets (1.3.1) is just a functor : $\Lambda \to$ Sets; while a direct system of abelian groups (1.3.11) is a functor : $\Lambda \to$ Abgp. (Exercise: verify these assertions.)

D. All the 'constructions' we have made are in fact functors:

Γ: Shfsp/X $\to$ Shv/X (2. 3. 4 and 2. 3. 7)

L: Presh/X $\to$ Shfsp/X (2. 3. 8 and 2. 3. 9).

Fix $x \in X$.

'stalk at x': Presh/X $\to$ Abgp : $F \in$ Ob(Presh) $\mapsto F_x$ (1. 5. 4).

We call the composite ΓL: Presh/X $\to$ Shv/X the sheafification functor.

E. In fact presheaves themselves can be considered as functors: if X is a topological space, let $\mathfrak{U}$ = set of all open subsets of X. $\mathfrak{U}$ is pre-ordered by the relation $\subseteq$ and so may be considered as a category (1. 4B). A presheaf of abelian groups on X is just a contravariant functor: $\mathfrak{U} \to$ Abgp (cf. 1. 1. 1).

We use this to formalise the generalisation mentioned in 1. 1. 2: if K is any category, a K-valued presheaf on X is a contravariant functor: $\mathfrak{U} \to$ K.

1. 7 **Definition.** If F, G : C $\to$ D are two functors, a natural transformation n from F to G is specified by giving for each $A \in$ ObC a morphism $n_A \in \mathrm{Hom}_D(FA, GA)$, in such a way that whenever $f \in \mathrm{Hom}_C(A, B)$ the square

$$\begin{array}{ccc} FA & \xrightarrow{n_A} & GA \\ {\scriptstyle F(f)}\downarrow & & \downarrow{\scriptstyle G(f)} \\ FB & \xrightarrow{n_B} & GB \end{array}$$

commutes (naturality) (or the square

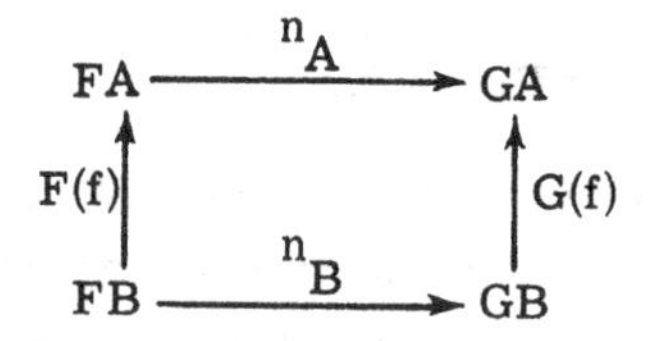

if F, G are contravariant). Natural transformations can be composed ($(n \circ m)_A = n_A \circ m_A$), and the functors F, G are called naturally equivalent (or naturally isomorphic) iff there are natural transformations n from F to G and m from G to F such that $n \circ m = id_G$ and $m \circ n = id_F$ (where id_F is the natural transformation with $(id_F)_A = id_{F(A)}$).

1.8 **Exercise.** Show that a natural transformation n is a natural equivalence iff $\forall A \in \text{Ob}C$ n_A is an isomorphism (cf. 1.5.2 and 1.9A below).

1.9 **Examples.**

A. If F, G are presheaves, considered as functors: $\mathfrak{U} \to$ Abgp as in 1.6E then a presheaf morphism $F \to G$ is just a natural transformation of functors (for the naturality square corresponding to $U \subseteq V$ is

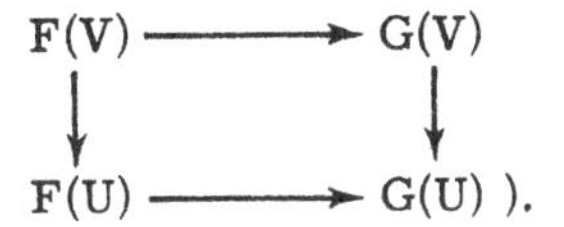

B. If Λ is a directed set, and $F : \Lambda \to$ Sets a direct system of sets as in 1.6C, then a target for F is just a natural transformation from F to a constant functor: $\Lambda \to$ Sets (Exercise: verify this).

C. 2.3.10 shows that the functor $L\Gamma$: Shfsp/X $\to$ Shfsp/X is naturally equivalent to the identity.

D. 2.4.1 and 2.4.4 show that there is a natural transformation n from the identity id: Presh $\to$ Presh to the sheafification functor ΓL: Presh $\to$ Shv $\hookrightarrow$ Presh; 2.4.3 shows that the sheafification functor $\Gamma L |$ Shv : Shv $\to$ Shv is naturally equivalent to id_{Shv}.

3.2 The categories of sheaves and presheaves

2.1 Our aim in Chapter 3 is to prove categorical properties of the categories Presh/X and Shv/X of presheaves and sheaves of abelian groups; in brief, to show that both are abelian categories and that the inclusion functor is left exact. We shall give definitions when needed, and refer to [Macl] and [Mit] for the wider view.

Analogous (but different) results could be proved for the categories of presheaves and sheaves of sets (see Exercises).

If $f \in \mathrm{Hom}_{\mathrm{Shv}}(F, G)$, in accordance with 2.5.7 we use the notation

$$\Gamma(U, f) : \Gamma(U, F) \to \Gamma(U, G)$$

for the map induced by f on the abelian groups of sections over an open set U. In this way

$$\Gamma(U, -) : \mathrm{Shv}/X \to \mathrm{Abgp}$$

becomes a functor, and we shall also be interested in its exactness properties.

2.2 If F, G are presheaves, given $f, g \in \mathrm{Hom}(F, G)$ we can construct a <u>sum</u> $(f + g) \in \mathrm{Hom}(F, G)$ by putting

$$(f + g)(U) : F(U) \to G(U) : s \mapsto f(s) + g(s)$$

(compatibility conditions are easily checked). There is a <u>zero morphism</u> $0 \in \mathrm{Hom}(F, G)$ with for each U

$$0(U) = \text{the zero map} : F(U) \to G(U) : s \mapsto 0.$$

We can verify easily that in this way Hom(F, G) becomes an abelian group, and that composition is bilinear, i. e. for a diagram of presheaves $F' \xrightarrow{p} F \underset{g}{\overset{f}{\rightrightarrows}} G \xrightarrow{q} G'$ we have

$$q \circ (f + g) = q \circ f + q \circ g$$

and

$$(f + g) \circ p = f \circ p + g \circ p.$$

2.3 We denote by 0 any zero sheaf on X, such that

$\forall$ open U $\quad 0(U)$ is a trivial abelian group

(e. g. the constant sheaf $\{0\}_X$ as in 2.4.6; this is the same as the constant presheaf $\{0\}_X$ of 1.2.A). For any presheaf, Hom(F, 0) and Hom(0, F) are each trivial groups. For

$f \in \mathrm{Hom}(F, G)$, clearly

$$f = 0 \iff f \text{ factors as } F \xrightarrow{f} G \text{ via } F \searrow 0 \nearrow G \quad .$$

2.4 2.2 and 2.3 show that both Presh and Shv are (pre-) <u>additive</u> categories (cf. [Macl] I§8, [Mit] I§18).

2.5 Exercise. Read enough about category theory to understand and justify the following concise summary of Chapters 1 and 2:

> If $\mathfrak{U}$ is the category of open sets of X, Presh is just the functor category $\mathrm{Abgp}^{\mathfrak{U}^{\mathrm{op}}}$ (op = dual). Shv is a full subcategory of Presh and we have a diagram of functors:
>
> 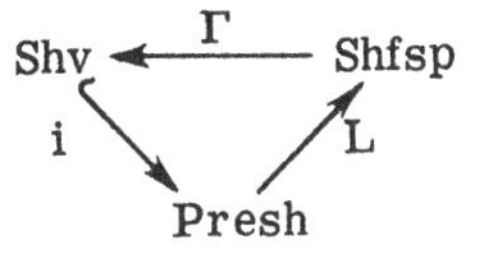
>
>
> in which Γ is an equivalence of categories (with inverse $L \circ i = L|\mathrm{Shv}$) and ΓL is left adjoint to the inclusion i (so that Shv is a reflective subcategory of Presh).

3.3 Kernels and monomorphisms

3.1 Definition. Given $f \in \mathrm{Hom}(F, G)$ for $F, G \in \mathrm{Ob}(\mathrm{Presh}/X)$, let

$$K(U) = \ker f(U) = \{s \in F(U);\ f(U)(s) = 0_{G(U)}\} \quad \text{(for } U \text{ open in } X)$$

a subgroup of $F(U)$. If $U \supseteq V$ are open in X and $s \in K(U)$ then

$$f(V)\rho^U_V(s) = \rho^U_V(f(U)(s)) = 0 \qquad \begin{array}{ccc} F(U) & \longrightarrow & G(U) \\ \downarrow & & \downarrow \\ F(V) & \longrightarrow & G(V) \end{array}$$

so $\rho^U_V(s) \in K(V)$.

Thus the $K(U)$ with the $\rho^U_V|K(U)$ form a presheaf over X, called the kernel of f and denoted Ker(f). We have a natural presheaf morphism $\mathrm{Ker}(f) \to F$ and the composite $\mathrm{Ker}(f) \to F \xrightarrow{f} G$ is zero.

3.2 Proposition. If $f \in \mathrm{Hom}(F, G)$, then Ker(f) has the universal property:

(*32) if H is a presheaf and $g \in \mathrm{Hom}(H, F)$ is such that $H \xrightarrow{g} F \xrightarrow{f} G = 0$, then g factors uniquely as (that is ↘ exists and is unique making the triangle commute).

$$\begin{array}{ccc} H & & \\ \downarrow & \searrow^{g} & \\ \mathrm{Ker}(f) & \longrightarrow & F \end{array}$$

Proof. Easy exercise. //

3.3 Proposition. If $f \in \mathrm{Hom}(F, G)$, F is a sheaf and G a monopresheaf, then Ker f is a sheaf. Hence if F, G are sheaves so is Ker f.

Proof. The monopresheaf condition is easy, since it works for F. For the glueing condition, if $U = \cup_{\lambda \in \Lambda} U_\lambda$ and $s_\lambda \in \mathrm{Ker}(f)(U_\lambda)$ satisfy the compatibility conditions on all $U_\lambda \cap U_\mu$, then $\exists s \in F(U)$ such that $\forall \lambda\ \rho^U_{U_\lambda}(s) = s_\lambda$, since F is a sheaf. Then $s' = f(U)(s) \in G(U)$ is such that $\forall \lambda\ \rho^U_{U_\lambda}(s') = 0$, and so $s' = 0$ since G is a monopresheaf. Hence $s \in \mathrm{Ker}(f)(U)$, as required. //

3.4 Remark. In an arbitrary category (with zero object) the universal property (*32) is used to define the notion of kernel (cf. [Macl] VIII §1, [Mit] I §13). 3.2 and 3.3 show that both the categories Presh and Shv have kernels of all morphisms, and the inclusion functor is kernel-preserving.

3.5 Theorem. For a morphism $f \in \mathrm{Hom}(F, G)$ of presheaves, the following three conditions are equivalent:

(i) $\mathrm{Ker}(f) = 0$

(ii) $\forall$open U in X f(U) is injective

(iii) f is a monomorphism i.e. if H is any presheaf and

$H \overset{g}{\underset{h}{\rightrightarrows}} F$ <u>are such that</u> $f \circ g = f \circ h$ <u>then</u> $g = h$.

<u>These imply the condition:</u>

(iv) $\forall x \in X$ f_x <u>is injective</u>

<u>which is a further equivalent condition if</u> F <u>is a monopresheaf.</u>

Proof. (i) $\Longleftrightarrow$ (ii) is immediate.

(i) $\Rightarrow$ (iii) Given $\mathrm{Ker}(f) = 0$ and H, g, h as in (iii), we see that $(g - h) : H \to F$ satisfies $H \xrightarrow{g-h} F \xrightarrow{f} G = 0$, so by Proposition 3.2 $(g - h)$ factors as $H \xrightarrow{g-h} F$ through 0 ($H \searrow 0 \nearrow F$), and so by 2.3 $(g - h) = 0$ i.e. $g = h$.

(iii) $\Rightarrow$ (i) If f is a monomorphism, we prove that $0 \to F$ has the universal property of $\mathrm{Ker}(f)$ stated in (*32): for if $H \xrightarrow{g} F \to G = 0$ then $H \overset{g}{\underset{0}{\rightrightarrows}} F \to G$ give the same result, so $g = 0$ by (iii) i.e. g factors (uniquely) as $H \xrightarrow{g} F$ through 0 ($H \searrow 0 \nearrow F$). Hence 0 has the universal property of $\mathrm{Ker}(f)$ and an easy argument like 1.3.6 shows that $\mathrm{Ker}(f)$ is a zero sheaf.

(ii) $\Rightarrow$ (iv) (F any presheaf.) Suppose $t \in F_x$ is such that $f_x(t) = 0$; then $\exists$ open U and $s \in F(U)$ such that s has germ t at x, so that $f(U)(s)$ has germ 0 at x, so that $\exists$ open $V \subseteq U$ with

$$0 = \rho^U_V(f(U)(s)) = f(V)(\rho^U_V(s)).$$

But $f(V)$ is injective by hypothesis, so $\rho^U_V(s) = 0$, and thus $t = 0$.

(iv) $\Rightarrow$ (ii) (assuming F a monopresheaf). Suppose $s \in F(U)$ is such that $f(U)(s) = 0 \in G(U)$; then $\forall x \in U$ $f_x(s_x) = (f(U)(s))_x = 0$, so $\forall x \in U$ $s_x = 0$ since each f_x is injective by hypothesis. Since F is a monopresheaf, $s = 0$ (cf. 2.3.1). //

3.6 **Definition.** If F, G are presheaves (resp. sheaves) on X and $f : F \to G$ is a monomorphism, we say that F is (or more precisely represents) a <u>subpresheaf</u> (resp. <u>subsheaf</u>) of G (via f). Two monomorphisms $F \rightarrowtail G$, $F' \rightarrowtail G$ are said to define the <u>same</u> sub(pre)sheaf of G iff $\exists$ an isomorphism $F \rightleftarrows F'$ of (pre)sheaves such that the diagram

$$\begin{array}{ccc} F & \searrow & \\ \downarrow\uparrow & & G \\ F' & \nearrow & \end{array}$$

commutes. If $G \xrightarrow{f} H$ is a morphism and

$\text{Ker } f \rightarrowtail G$ and $F \rightarrowtail G$ represent the same subpresheaf of G in this sense, then we see easily that $F \to G$ has the universal property of a kernel stated in 3.2, and we say that $F \to G$ is _a kernel_ of $G \to H$.

The definitions of monomorphism and of subobject given here make sense in an arbitrary category (cf. [Macl] I §5, V §7; [Mit] I §5).

3.7 Corollary. _If_ F _is a monopresheaf, then it is a subpresheaf of a sheaf (namely its sheafification); conversely a subpresheaf of a sheaf is always a monopresheaf._

Proof. 2.4.5 and (iii) $\Leftrightarrow$ (iv) of 3.5; the converse is easy directly from 3.5(ii) and 2.1.1. //

3.8 Corollary. _If_ $f : E \to E'$ _is a morphism of sheaf spaces over_ X, _then_ $\Gamma f : \Gamma E \to \Gamma E'$ _is a monomorphism of sheaves_ $\Leftrightarrow$ f _is injective_ $\Leftrightarrow$ f _is a homeomorphism of_ E _onto an open subspace of_ E'.

Proof. 2.3.6 and 2.3.7 and (iii) $\Leftrightarrow$ (iv) of 3.5; then 2.3.5(c). //

3.9 Proposition. _If_ $f : F \to G$ _is a morphism of presheaves, then_

$$(\text{Ker } f)_x = \text{Ker } f_x$$

(_equality as subgroups of_ F_x, _which the LHS is by_ (ii) $\Rightarrow$ (iv) _of 3.5, since_ $\text{Ker } f \to F$ _is a monomorphism_).

Proof.

$$\begin{aligned} t \in (\text{Ker } f)_x &\Leftrightarrow \exists \text{ open } U \ni x \text{ and } s \in \text{Ker}(f)(U) \text{ such that } t = s_x \\ &\Leftrightarrow \exists \text{ open } U \ni x \text{ and } s \in F(U) \text{ such that } t = s_x \text{ and } \\ &\qquad\qquad f(U)(s) = 0 \\ &\Leftrightarrow f_x(t) = 0. \ // \end{aligned}$$

3.10 Definition. If F, F' are subpresheaves of G, we write $F \le F'$ iff $\exists$ a morphism $F \to F'$ (necessarily mono) such that the triangle $F \to F'$, $F \searrow G$, $F' \swarrow G$ commutes. Hence $F \le F' \le F \Leftrightarrow F$, F' are the same subpresheaf (3.6).

3.11 Proposition. *If* F, F' *are subsheaves of a sheaf* G, *then*

$$F \le F' \iff \forall x \in X \;\; F_x \subseteq F'_x$$

(as subgroups of G_x *by 3.5(iv)).*

Proof. $\Rightarrow$: $F \le F' \Rightarrow \exists\ \begin{array}{ccc} F & \xrightarrow{\text{mono}} & F' \\ & \searrow \;\; \swarrow & \\ & G & \end{array} \Rightarrow \forall x \in X\ \exists\ \begin{array}{ccc} F_x & \xrightarrow{\text{inj}} & F'_x \\ & \searrow \;\; \swarrow & \\ & G_x & \end{array}$

i.e. $F_x \subseteq F'_x$.

$\Leftarrow$: Suppose $\forall x \in X$ $F_x \subseteq F'_x$. Given $s \in \Gamma(U, F)$ with U open in X, there is a unique section of F' with germs s_x at each $x \in U$; hence there is a morphism : $F \to F'$ such that the maps $F \to F' \to G$ and $F \to G$ agree on all stalks. By 2.1.10 they are equal and so $F \le F'$. //

3.12 Corollary. *For two subsheaves* F, F' *of a sheaf* G *we have*

$$F = F' \iff \forall x \in X \;\; F_x = F'_x.$$

Proof. Apply 3.11 twice. //

3.4 Cokernels and epimorphisms

4.1 Definition. Given $f \in \mathrm{Hom}(F, G)$ for $F, G \in \mathrm{Ob}(\mathrm{Presh}/X)$ let

$$C(U) = G(U)/f(U)[F(U)] = G(U)/\mathrm{Im}\, f(U) \quad (\text{for } U \text{ open in } X)$$

($\mathrm{Im}\, f(U)$ is a sub-abelian group of $G(U)$ - take the quotient group). If $U \supseteq V$ are open in X, the map $\rho^U_V : G(U) \to G(V)/\mathrm{Im}\, f(V)$ kills $\mathrm{Im}\, f(U)$: for if $s \in F(U)$ then

$$\rho^U_V(f(U)(s)) = f(V)\rho^U_V(s) \in \mathrm{Im}\, f(V).$$

Hence we get an induced $\bar{\rho}^U_V : C(U) \to C(V)$ and C becomes a presheaf, called the *presheaf cokernel* of f and denoted by $\mathrm{PCok}(f)$. We have a natural presheaf morphism: $G \to \mathrm{PCok}(f)$ and the composite $F \to G \to \mathrm{PCok}(f)$ is zero.

4.2 Proposition. <u>If</u> $f \in \mathrm{Hom}(F, G)$, <u>then</u> PCok(f) <u>has the universal property:</u>

(*42) <u>if</u> H <u>is a presheaf and</u> $g \in \mathrm{Hom}(G, H)$ <u>is such that</u> $F \xrightarrow{f} G \xrightarrow{g} H = 0$ <u>then</u> g <u>factors uniquely as</u> (that is there is one and only one morphism ↗ making the triangle commute).

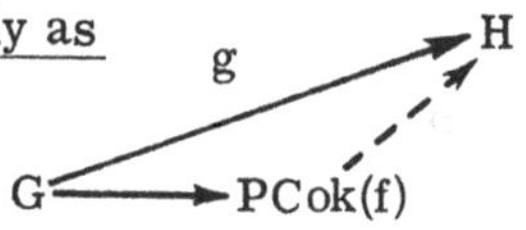

$$\begin{array}{ccc} & & H \\ & \overset{g}{\nearrow} & \uparrow \\ G & \longrightarrow & \mathrm{PCok}(f) \end{array}$$

Proof. Easy exercise. //

4.3 Warning. If F, G are sheaves and $f \in \mathrm{Hom}(F, G)$, then PCok(f) need not be a sheaf (cf. 4.9).

4.4 Definition. If F, G are sheaves on X and $f \in \mathrm{Hom}(F, G)$, then the <u>sheaf cokernel</u> of f SCok(f) is the sheafification (2.4.1) ΓLPCok(f) of the presheaf cokernel. Thus SCok(f) is a sheaf, and we have a natural morphism $G \to \mathrm{SCok}(f)$ (given by $G \to \mathrm{PCok}(f) \to \mathrm{SCok}(f)$) such that the composite $F \to G \to \mathrm{SCok}(f)$ is zero.

4.5 Proposition. <u>If</u> $f \in \mathrm{Hom}(F, G)$ <u>with</u> F, G <u>sheaves, then</u> SCok(f) <u>has the universal property:</u>

(*45) <u>if</u> H <u>is a sheaf and</u> $g \in \mathrm{Hom}(G, H)$ <u>is such that</u> $F \xrightarrow{f} G \xrightarrow{g} H = 0$, <u>then</u> g <u>factors uniquely as</u>

$$\begin{array}{ccc} & & H \\ & \overset{g}{\nearrow} & \uparrow \\ G & \longrightarrow & \mathrm{SCok}(f) \end{array}$$

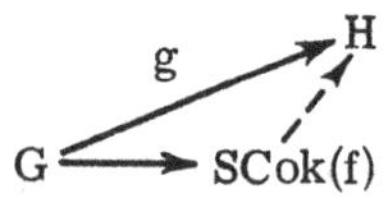

Proof. We get unique factorisation of g by presheaf morphisms through PCok(f) by Proposition 4.2; thus through SCok(f) by Theorem 2.4.2, since H is a sheaf. //

4.6 Remark. The universal properties (*42) and (*45) are the versions in Presh and Shv of the definition of a cokernel in an arbitrary category (with zero object) (cf. [Macl] III §3, VIII §1, [Mit] I, §13). Thus both Presh and Shv have cokernels of all morphisms, but the inclusion functor is not cokernel-preserving (cf. 4. 3 and 4. 9).

4. 7 **Theorem.** Let F, G be presheaves and $f \in \mathrm{Hom}(F, G)$. Then the following three conditions are equivalent:

(i) $\mathrm{PCok}(f) = 0$

(ii) $\forall$ open U in X $f(U)$ is surjective

(iii) f is an epimorphism in Presh i. e. if H is any presheaf and $G \underset{h}{\overset{g}{\rightrightarrows}} H$ are such that $g \circ f = h \circ f$, then $h = g$.

Proof. (i) $\Longleftrightarrow$ (ii) is immediate.

(i) $\Longleftrightarrow$ (iii) is easy (apply 4. 2 to $(g - h)$; cf. 3. 5). //

4. 8 **Theorem.** Let $f : F \to G$ be a morphism of sheaves. Then the following four conditions are equivalent:

(i) $\mathrm{SCok}(f) = 0$

(ii) $\forall x \in X \quad (\mathrm{PCok}(f))_x = 0$

(iii) $\forall x \in X \quad f_x$ is surjective

(iv) f is an epimorphism in Shv i. e. if H is any sheaf and $G \underset{h}{\overset{g}{\rightrightarrows}} H$ are such that $g \circ f = h \circ f$ then $g = h$.

Furthermore any of the conditions of 4. 7 implies all of these.

Proof. (i) $\Longleftrightarrow$ (iv) is easy (apply 4. 5 to $(g - h)$).

(i) $\Longleftrightarrow$ (ii) $\mathrm{SCok}(f) = 0 \Longleftrightarrow \forall x \in X \quad (\mathrm{SCok}(f))_x = 0$ since $\mathrm{SCok}(f)$ is a sheaf (apply 2. 1. 10 to the morphisms $\mathrm{SCok}(f) \underset{0}{\overset{id}{\rightrightarrows}} \mathrm{SCok}(f)$).

$\Longleftrightarrow \forall x \in X \quad (\mathrm{PCok}(f))_x = 0$ by 2. 4. 5

(ii) $\Longleftrightarrow$ (iii) $\mathrm{PCok}(f)_x = 0 \Longleftrightarrow \forall$ open $U \ni x$ and $s \in \mathrm{PCok}(f)(U)$ $\exists$ open V with $U \supseteq V \ni x$ and $\rho^U_V(s) = 0$

$\Longleftrightarrow \forall$ open $U \ni x$ and $s \in G(U)$ $\exists$ open V with $U \supseteq V \ni x$ and $\rho^U_V(s) \in \mathrm{Im} f(V)$

$\Longleftrightarrow f_x$ surjective.

Finally, clearly 4. 7(iii) $\Rightarrow$ 4. 8(iv) (aliter 4. 7(i) $\Rightarrow$ 4. 8(ii)). //

4. 9 **Examples** to show the inequivalence of the conditions of 4. 7 and 4. 8 and to justify 4. 3 and 4. 6.

A. Let $X = [0, 1] \hookrightarrow \mathbf{R}$ and let F be the constant sheaf with stalks $\mathbf{Z}$ (with sheaf space $X \times \mathbf{Z} \xrightarrow{\pi_1} X$). Let G be the sheaf whose stalks are

$$G_x = \begin{cases} \mathbf{Z} & \text{if } x = 0 \text{ or } 1 \\ 0 & \text{otherwise} \end{cases}$$

so that, for instance, $\Gamma(X, G) = \mathbf{Z} \oplus \mathbf{Z}$. Let $f : F \to G$ be the unique morphism such that $f_x = id_{\mathbf{Z}}$ if $x = 0$ or 1. Then f is clearly a sheaf epimorphism, by 4.8(iii); but

$$\Gamma(X, f) : \mathbf{Z} \to \mathbf{Z} \oplus \mathbf{Z}$$

cannot be surjective, so by 4.7(ii) f is not a presheaf epimorphism.

B. Let $X = \mathbf{C}$ and let C^ω be the abelian sheaf of $\mathbf{C}$-valued analytic functions on $\mathbf{C}$. Let $d : C^\omega \to C^\omega$ be the morphism of differentiation:

$$d(U) : C^\omega(U) \to C^\omega(U) : f \mapsto \frac{df}{dz} \quad \text{for } U \text{ open in } \mathbf{C}.$$

For any $x \in \mathbf{C}$, an analytic function on a small disc neighbourhood of x can be 'integrated' and expressed as $\frac{df}{dz}$ with f analytic (look at the power series expansion); hence d is a sheaf epimorphism (4.8(iii)). But if we take U to be not simply connected, there are analytic functions on U which cannot be expressed as $d(U)$ of an analytic function e.g. $\frac{1}{z}$ on $z \neq 0$; hence such $d(U)$ are not surjections and d is not a presheaf epimorphism.

4.10 **Theorem.** For $f : F \to G$ a morphism of presheaves, we have the following equivalent conditions:

(i) f is an isomorphism;

(ii) $\forall$ open U of X $f(U)$ is bijective;

(iii) f is a monomorphism and a presheaf epimorphism.

If f is a morphism of sheaves, we have the further equivalent conditions:

(iv) f is a monomorphism and a sheaf epimorphism;

(v) $\forall x \in X$ f_x is bijective.

Proof. (i) $\Longleftrightarrow$ (ii) is 1.5.2.

(ii) $\Longleftrightarrow$ (iii) is clear from 3.5 and 4.7.

Now suppose that F, G are sheaves.

(iv) $\Longleftrightarrow$ (v) is 3.5(iv) and 4.8(iii).

(iii) $\Rightarrow$ (iv) is 4.7 $\Rightarrow$ 4.8.

(v) $\Rightarrow$ (i): If (v) is true, then by the Construction 2.3.8 the morphism of sheaf spaces $Lf : LF \to LG$ is bijective, and it is a local homeomorphism by 2.3.5(c); hence it is an isomorphism of sheaf spaces, and so in

$$\begin{array}{ccc} \Gamma LF & \xrightarrow{\Gamma Lf} & \Gamma LG \\ \wr\downarrow & & \downarrow\wr \\ F & \xrightarrow{f} & G \end{array}$$

(which commutes by 2.4.4) ΓLf is an isomorphism.
Hence f is too by 2.4.3. //

4.11 **Proposition.** Let $f : F \to G$ be a morphism of presheaves. Then

$$\forall x \in X \quad (PCok\ f)_x = Cok\ f_x\ (= G_x/Im\ f_x),$$

(equality as quotient abelian groups of G_x, which LHS is by 4.8(iii) since $G \to PCok(f)$ is a presheaf epimorphism). If f is a morphism of sheaves, then in addition

$$\forall x \in X \quad (SCok\ f)_x = Cok\ f_x.$$

Proof. For $t \in G_x$ we have

$$\begin{aligned} t \mapsto 0 \in (PCok(f))_x &\Longleftrightarrow \exists \text{ open } U \ni x \text{ and } s \in G(U) \text{ such that } s_x = t \text{ and} \\ &\qquad s \mapsto 0 \in PCok(f)(U) \\ &\Longleftrightarrow \exists \text{ open } U \ni x \text{ and } s \in G(U) \text{ such that } s_x = t \text{ and} \\ &\qquad s \in Im\ f(U) \\ &\Longleftrightarrow t \in Im(f_x). \end{aligned}$$

In the sheaf case, we have a commutative diagram G → $PCok(f)$, G → $SCok(f)$, $PCok(f) \downarrow SCok(f)$ inducing on stalks

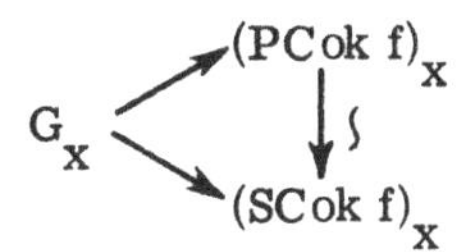

where the vertical arrow is isomorphic by 2.4.5. Hence the result. //

4.12 **Definition.** Dually to 3.6, if $f : F \to G$ is a presheaf epimorphism, we say that G is a *quotient presheaf* of F. If F, G are sheaves and f is a sheaf epimorphism, we say that G is a *quotient sheaf* of F. In either case

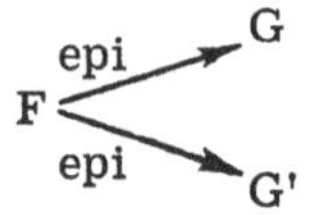

define the *same* quotient object of F iff there is an isomorphism $G \xrightarrow{\sim} G'$ such that $F \rightrightarrows$ ($F \to G$, $F \to G'$, $G \xrightarrow{\wr} G'$) commutes. As in 3.6 we can define the notions of a presheaf cokernel and a sheaf cokernel.

The definitions of epimorphism and quotient object given here make sense in an arbitrary category (cf. references for 3.6).

4.13 **Theorem.** (i) *Every subpresheaf is a kernel of some presheaf morphism.*

(ii) *Every subsheaf is a kernel of some sheaf morphism.*

(iii) *Every quotient presheaf is a (presheaf) cokernel of some presheaf morphism.*

(iv) *Every quotient sheaf is a sheaf cokernel of some sheaf morphism.*

In other words, both Presh and Shv are categories satisfying:

(a) *every monomorphism is a kernel, and*

(b) *every epimorphism is a cokernel.*

Proof. (i) Given $f : F \overset{\text{mono}}{\to} G$ in Presh, we wish to prove $F = \text{Ker}(G \to \text{PCok}\, f)$. But $G \to \text{PCok}(f)$ kills (sends to zero) exactly those sections $s \in G(U)$ which arise by f from sections of $F(U)$.

(ii) Given $f : F \overset{\text{mono}}{\to} G$ in Shv, we have $F \le \text{Ker}(G \to \text{SCok}(f))$ (as subobjects of G); for by the universal property of Ker (3.2), since $F \to G \to \text{SCok}\, f = 0$, $F \to G$ factors through the kernel. But for $x \in X$

$$\begin{aligned} F_x &= \mathrm{Ker}(G_x \to (\mathrm{SCok}\ f)_x) && \text{by } 4.11 \\ &= (\mathrm{Ker}(G \to \mathrm{SCok}\ f))_x && \text{by } 3.9 \end{aligned}$$

and so $F = \mathrm{Ker}(G \to \mathrm{SCok}\ f)$ by 3.11.

(iii) Given $f : F \overset{\text{epi}}{\to} G$ in Presh, we wish to prove $G = \mathrm{PCok}(\mathrm{Ker}\ f \to F)$, as quotient presheaves of F. By 4.2 we get

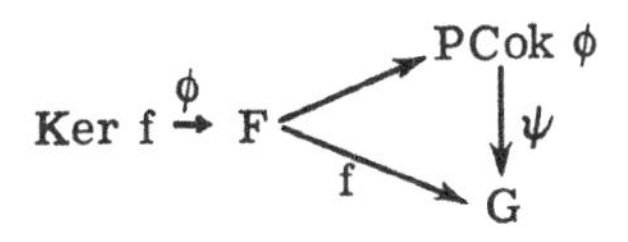

and ψ is a presheaf epimorphism since f is, and ψ is easily seen to be monomorphic, so ψ is an isomorphism by 4.10.

(iv) Given $f : F \overset{\text{epi}}{\to} G$ in Shv, we wish to prove $G = \mathrm{SCok}(\mathrm{Ker} f \to F)$ as quotient sheaves of F; by 4.5 we get

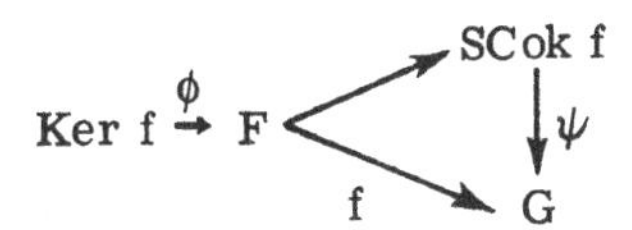

and $\forall x \in X$ ψ_x is surjective since f_x is (4.8), and $\forall x \in X$ ψ_x is injective by 4.11. Hence by 4.10 ψ is an isomorphism of sheaves. //

4.14 **Remark.** Given a subobject $F \to G$, we are tempted, by analogy with Abgp, to write G/F for the quotient object of G whose kernel is F, as in 4.13(i) and (ii). But this notation does not lend itself well to the distinction between the categories Presh and Shv; as 4.9 shows, G/F may differ in the two.

3.5 Biproducts, and the abelianness of Presh and Shv

5.1 **Construction.** Let F, G be presheaves of abelian groups on a topological space X. We can define a presheaf $F \oplus G$ (the direct sum, or biproduct of F and G) by

$$(F \oplus G)(U) = F(U) \oplus G(U) \qquad \text{for } U \text{ open in } X$$

$$\rho_V^U : F(U) \oplus G(U) \to F(V) \oplus G(V) : (s, t) \mapsto (\rho_V^U(s), \rho'^U_V(t)) \qquad \text{for } U \supseteq V \text{ open in } X$$

and we have natural presheaf morphisms:

$$\begin{array}{ccccc} F & \xrightarrow{\amalg_1} & F \oplus G & \xrightarrow{\pi_1} & F \\ G & \xrightarrow{\amalg_2} & F \oplus G & \xrightarrow{\pi_2} & G \end{array}$$

where for example $\amalg_1(s) = (s, 0)$ for $s \in F(U)$

$\pi_2(s, t) = t$ for $s \in F(U)$, $t \in G(U)$.

5.2 **Proposition.** $F \oplus G$ has the universal property of a biproduct in Presh: namely

(i) for any presheaf H and morphisms $F \to H$, $G \to H$, there is a unique morphism $F \oplus G \to H$ such that

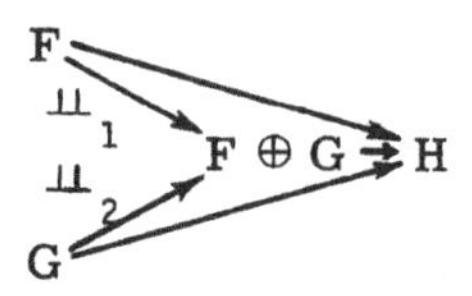

commutes (that is, $F \oplus G$ is a categorical sum (coproduct) of F and G).

(ii) for any presheaf H and morphisms $H \to F$, $H \to G$, there is a unique morphism $H \to F \oplus G$ such that

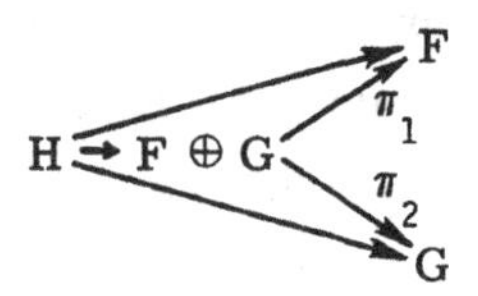

commutes (that is, $F \oplus G$ is a categorical product of F and G).

Proof. Easy, using the same result for the category Abgp. //

5.3 **Proposition.** If F and G are sheaves, then so is $F \oplus G$; hence in this case $F \oplus G$ has the universal property of a biproduct in Shv (so that 5.2 holds with H always a sheaf).

Proof. The exactness of

$$0 \to H(U) \to \Pi_{\lambda \in \Lambda} H(U_\lambda) \to \Pi_{(\lambda, \mu) \in \Lambda \times \Lambda} H(U_\lambda \cap U_\mu)$$

for $H = F$, G implies its exactness for $H = F \oplus G$. //

5.4 **Remark.** Hence both Presh and Shv have biproducts (cf. [Macl] VIII §2, [Mit] I §§17, 18), and the inclusion-functor is biproduct-preserving. Recall that a category C is said to be abelian iff

(a) C is additive: that is, C has a zero object and each set $\mathrm{Hom}_C(A, B)$ has a natural abelian group structure (as in 3.2.2);

(b) C has a biproduct of each pair of objects;

(c) C has kernels and cokernels of all morphisms;

(d) in C every monomorphism is a kernel, and every epimorphism a cokernel.

(cf. [Macl] VIII §3, [Mit] I §20.)

5.5 **Theorem.** Both Presh/X and Shv/X are abelian categories, and the inclusion functor Shv ↪ Presh is kernel-preserving.

Proof. 3.2.2, 3.2.3; 5.2 and 5.3; 3.2 and 3.3; 4.2 and 4.5; 4.13; 3.3. //

3.6 **Exact sequences**

6.1 **Definition.** (i) If $f : F \to G$ is a morphism of presheaves, we define the (presheaf) image of f to be $\mathrm{PIm}(f) = \mathrm{Ker}(G \to \mathrm{PCok}\, f)$.

(ii) If $f : F \to G$ is a morphism of sheaves, we define the (sheaf) image of f to be $\mathrm{SIm}(f) = \mathrm{Ker}(G \to \mathrm{SCok}\, f)$.

6.2 **Exercise.** Formulate the universal property that you would like a concept of 'image' to satisfy, and verify that PIm and SIm do in the categories Presh and Shv.

6.3 **Exercise.** Check that PIm(f) is a presheaf whose abelian group of sections over each open U is the image of f(U), while SIm(f) is a sheaf whose stalk at each $x \in X$ is the image of f_x.

6.4 **Definition.** Let $\ldots \to F \xrightarrow{f} G \xrightarrow{g} H \to \ldots$ be a sequence of presheaves and morphisms over a space X. We say that the sequence is exact at G as a sequence of presheaves iff

$$\mathrm{PIm}(f) = \mathrm{Ker}(g) \quad \text{(equal subpresheaves of } G);$$

and that it is an exact sequence of presheaves iff it is exact at each point at which this condition makes sense.

If the sequence consists of sheaves, we say that it is exact at G as a sequence of sheaves iff

$$\mathrm{SIm}(f) = \mathrm{Ker}(g) \quad \text{(equal subsheaves of } G);$$

and define an exact sequence of sheaves analogously.

These definitions make sense in any category with kernels and cokernels, although they are primarily used in abelian categories. In particular we shall use the concept of exactness in Abgp.

6.5 **Theorem.** (i) $F \to G \to H$ is an exact sequence of presheaves iff $\forall$ open U in X $F(U) \to G(U) \to H(U)$ is an exact sequence of abelian groups.

(ii) $F \to G \to H$ is an exact sequence of sheaves iff $\forall x \in X$ $F_x \to G_x \to H_x$ is an exact sequence of abelian groups.

(iii) If $F \xrightarrow{f} G \xrightarrow{g} H$ is a sequence of sheaves which is exact as a sequence of presheaves, then it is an exact sequence of sheaves.

Proof.

(i) $F \xrightarrow{f} G \xrightarrow{g} H$ exact $\iff$ $\mathrm{Ker}\, g = \mathrm{PIm}\, f$

$\iff \forall$ open U $\mathrm{Ker}\, g(U) = \mathrm{Ker}(G(U) \to G(U)/\mathrm{Im} f(U))$

$= \mathrm{Im}\, f(U)$

$\iff \forall$ open U $F(U) \to G(U) \to H(U)$ exact.

(ii) $F \xrightarrow{f} G \xrightarrow{g} H$ exact $\iff \forall x \in X$ $(\mathrm{Ker}\, g)_x = (\mathrm{SIm}\, f)_x$ by 3.12.

But $(\mathrm{SIm}\, f)_x = (\mathrm{Ker}(G \to \mathrm{SCok}(f)))_x$ by definition

$= \mathrm{Ker}(G_x \to (\mathrm{SCok}\, f)_x)$ by 3.9

$= \mathrm{Ker}(G_x \to G_x/\mathrm{Im}\, f_x)$ by 4.11

$= \mathrm{Im}\, f_x$.

Hence the result.

(iii) We have $\operatorname{Ker} g = \operatorname{PIm} f$; hence $\forall x \in X$

$$(\operatorname{Ker} g)_x = (\operatorname{PIm} f)_x = \operatorname{Ker}(G_x \to (\operatorname{PCok} f)_x) = \operatorname{Ker}(G_x \to (\operatorname{SCok} f)_x) \text{ by 4.11}$$
$$= (\operatorname{SIm} f)_x$$

and so $\operatorname{Ker} g = \operatorname{SIm} f$ by 3.12. //

6.6 **Corollary.** In each of the categories Presh and Shv:

(a) $0 \to F \xrightarrow{f} G$ is exact $\iff$ f is a monomorphism.

(b) $F \xrightarrow{f} G \to 0$ is exact $\iff$ f is an epimorphism.

(c) For any morphism $f : F \to G$

$$0 \to \operatorname{Ker} f \to F \to G \to {}^{P}_{S}\operatorname{Cok}(f) \to 0$$

is exact.

(d) $0 \to F \xrightarrow{f} G \xrightarrow{g} H$ is exact $\iff$ f is a kernel of g.

(e) $F \xrightarrow{f} G \xrightarrow{g} H \to 0$ is exact $\iff$ g is a cokernel of f.

Proof. Direct from 6.5, using 3.5, 4.7 and 4.8, 3.9 and 4.11. Aliter: this result holds in any abelian category, and may be proved by using the universal properties of monomorphisms, epimorphisms, kernels and cokernels. //

6.7 **Definition.** A covariant functor T between two categories in each of which the concept of exactness is defined (e.g. two abelian categories) is called exact (resp. left exact, right exact) iff whenever

$$0 \to F \to G \to H \to 0$$

is an exact sequence, the sequence

$$0 \to TF \to TG \to TH \to 0$$

(resp. $0 \to TF \to TG \to TH$, $TF \to TG \to TH \to 0$) is also exact.

6.8 **Exercise.** Prove that T is exact $\iff$ T preserves all exact sequences.

6.9 **Theorem.** (i) The inclusion functor Shv $\to$ Presh is left exact.

(ii) The functor 'sheafify' $= \Gamma L$: Presh $\to$ Shv is exact.

(iii) For each open U in X the functor 'sections over U'

$$-(U) : \text{Presh} \to \text{Abgp} : F \mapsto F(U)$$

is exact.

(iv) For each open U in X the functor 'sections over U'

$$\Gamma(U, -) : \text{Shv} \to \text{Abgp} : F \mapsto \Gamma(U, F)$$

is left exact.

Proof. (i) If $0 \to F \xrightarrow{f} G \xrightarrow{g} H \to 0$ is exact in Shv, then f is a kernel of g, and so $0 \to F \to G \to H$ is exact in Presh (all by 6.6).

(ii) An exact sequence $0 \to F \to G \to H \to 0$ of presheaves over X gives, for each $x \in X$ an exact stalk sequence $0 \to F_x \to G_x \to H_x \to 0$ (by, for instance 3.5 and 4.11, with 6.6(a) and (e) in mind). By 2.4.4 and 2.4.5 there is a commutative diagram

$$\begin{array}{ccccccccc} 0 & \to & F_x & \to & G_x & \to & H_x & \to & 0 \\ & & \downarrow\wr & & \downarrow\wr & & \downarrow\wr & & \\ 0 & \to & (\Gamma LF)_x & \to & (\Gamma LG)_x & \to & (\Gamma LH)_x & \to & 0 \end{array}$$

with the vertical arrows isomorphisms. Hence the lower sequence is also exact, and so $0 \to \Gamma LF \to \Gamma LG \to \Gamma LH \to 0$ is an exact sequence of sheaves, by 6.5(ii).

(iii) Is a restatement of 6.5(i).

(iv) If $0 \to F \xrightarrow{f} G \xrightarrow{g} H \to 0$ is exact, then f is a kernel of g and so for each open U, $\Gamma(U, f)$ is a kernel of $\Gamma(U, g)$; that is the sequence $0 \to \Gamma(U, F) \to \Gamma(U, G) \to \Gamma(U, H)$ is exact in Abgp. //

6.10 **Example.** Continuing the example of 4.9B, we have an exact sequence of sheaves over $X = \mathbf{C}$

$$(*) \qquad 0 \to \mathbf{C} \to C^{\omega} \xrightarrow{d = \text{differentiate}} C^{\omega} \to 0$$

where C denotes the constant sheaf on X. Thus for any connected open set U in X

$$0 \to C \to \Gamma(U, C^{\omega}) \xrightarrow{\Gamma(U, d)} \Gamma(U, C^{\omega})$$

is an exact sequence of abelian groups; but the last arrow need not be surjective, as in 4.9B (which shows that (*) is not an exact sequence of presheaves). We shall see later in our study of cohomology that this failure of right exactness can be regarded as a property of the kernel (viz C) as a sheaf on U.

6.11 **Remark.** Some authors succumb to the temptation to define exactness of sequences of presheaves and of sheaves by the properties 6.5(i) and (ii); but this really begs the question of whether this concept of exactness is the usual one in an abelian category, as we have defined it. However, once 6.5 is known, it is a convenient criterion.

3.7 Change of base space

7.1 **Construction.** Suppose we are given a continuous map $\phi : X \to Y$ of topological spaces, and a presheaf F on X. We obtain a presheaf $\phi_* F$ on Y, called the direct image of F by ϕ, by putting

$$\begin{cases} (\phi_* F)(U) = F(\phi^{-1} U) \\ \rho^U_V \qquad = \rho^{\phi^{-1} U}_{\phi^{-1} V} \end{cases} \qquad \text{for } U \supseteq V \text{ open in } Y.$$

If $f : F \to G$ is a morphism of presheaves on X, we get a morphism $\phi_* f : \phi_* F \to \phi_* G$ in an obvious way (over an open U, $\phi_* f(U) = f(\phi^{-1} U)$).

7.2 **Remark.** If we replace X, Y by their categories of open sets $\mathcal{U}$, $\mathcal{V}$ (as in 3.1.6E), then the continuous map ϕ can be regarded as a functor: $\mathcal{V} \to \mathcal{U}$. Regarding F as a functor: $\mathcal{U} \to$ Abgp (loc. cit.) the functor $\phi_* F$ is just the composite $\mathcal{V} \to \mathcal{U} \to$ Abgp.

7.3 **Proposition.** If F is a sheaf on X, then so is $\phi_* F$ on Y.

Proof. Straightforward: reduce the properties (M) and (G) of

2.1.1 and 2.1.3 for a cover (U_i) of $U \subseteq Y$ to the same properties for the cover $(\phi^{-1}(U_i))$ of $\phi^{-1}(U) \subseteq X$. //

7.4 **Exercise.** Verify the functorial properties of ϕ_*:

$$\phi_*(id_F) = id_{\phi_* F}; \quad \phi_*(f \circ g) = \phi_* f \circ \phi_* g.$$

Verify the functorial properties of $-_*$: $id_* = id$, $(\phi \circ \psi)_* = \phi_* \circ \psi_*$. Hence given a continuous map $\phi : X \to Y$, ϕ_* is a functor: Presh/X $\to$ Presh/Y, and by 7.3 restricts to a functor ϕ_* : Shv/X $\to$ Shv/Y.

7.5 **Examples.** A. For any X, let Y have one point and ϕ be the only map. Then Shv/Y $\cong$ Abgp are isomorphic categories, and ϕ_* : Shv/X $\to$ Shv/Y $\cong$ Abgp is the same functor as $\Gamma(X, -)$.

B. Let Y be a nice space, such as $\mathbf{R}^2$, and X a nice open subset, such as an open disc. Let $\phi : X \hookrightarrow Y$ and F be a constant sheaf on X, say with stalks $\mathbf{Z}$.

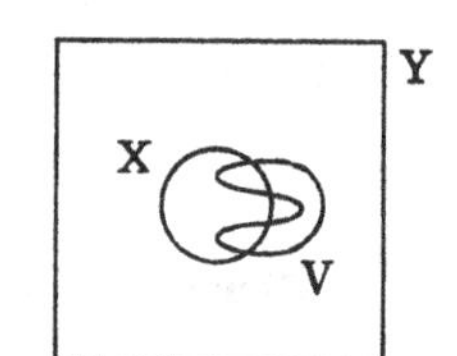

Let G be the constant sheaf $\mathbf{Z}$ on Y. Then we see that for the open set V illustrated

$$\Gamma(V, \phi_* F) = \mathbf{Z} \oplus \mathbf{Z}, \text{ while } \Gamma(V, G) = \mathbf{Z}.$$

Hence $\phi_* F$ is not a constant sheaf on Y. In fact the stalks of $\phi_* F$ are easily seen to be given by

$$(\phi_* F)_x = \begin{cases} \mathbf{Z} & \text{if } x \in \overline{X} \\ 0 & \text{otherwise} \end{cases}$$

so that ϕ_* 'spreads' F out onto the closure of X.

C. Let $X = Y = S^1$ be the circle, and $\phi : X \to Y$ a double covering (given by $z \mapsto z^2$ if we represent S^1 as $\{z \in \mathbf{C}; |z| = 1\}$). Let F be the constant sheaf $\mathbf{Z}$ on X.

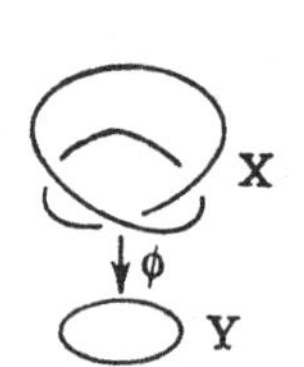

Then if V is a small open interval in Y, we have

$$\phi_* F(V) = F(\phi^{-1}(V)) = \mathbf{Z} \oplus \mathbf{Z}$$

so that the stalks of $\phi_* F$ are

$$(\phi_* F)_y = \mathbf{Z} \oplus \mathbf{Z} \text{ for each } y \in Y.$$

But $\phi_* F$ is not a constant sheaf, since

$$\Gamma(Y, \phi_* F) = \Gamma(X, F) = \mathbf{Z}.$$

[Note that this gives an example of two non-isomorphic sheaves which nevertheless have the same stalks; compare 3.12.]

7.6 **Proposition.** The functor $\phi_* : \text{Shv}/X \to \text{Shv}/Y$ is left exact.

Proof. If $0 \to F \to G \to H \to 0$ is exact in Shv/X, then for V open in Y, let $U = \phi^{-1}V$. By 6.9(iv)

$$0 \to \Gamma(U, F) \to \Gamma(U, G) \to \Gamma(U, H) \text{ is exact in Abgp},$$

that is

$$0 \to \Gamma(V, \phi_* F) \to \Gamma(V, \phi_* G) \to \Gamma(V, \phi_* H) \text{ is exact in Abgp},$$

so that $0 \to \phi_* F \to \phi_* G \to \phi_* H$ is exact in Shv/Y by 6.5(i) and (iii). //

7.7 **Remark.** After example 7.5A we see that the left exactness of ϕ_* is a generalisation of the left exactness of $\Gamma(X, -) : \text{Shv}/X \to \text{Abgp}$ (6.9(iv)).

7.8 **Definition.** If $\phi : X \to Y$ is a continuous map and F, G are presheaves on X, Y respectively, a morphism $f : G \to F$ over ϕ (or ϕ-morphism) is given by a collection of Abgp morphisms

$$f(U, V) : G(V) \to F(U)$$

for all U open in X, V open in Y such that $U \subseteq \phi^{-1}V$, subject to the condition that if $U \supseteq U'$, $V \supseteq V'$ with U', V' open and such that $U' \subseteq \phi^{-1}(V')$ then the square commutes:

$$\begin{array}{ccc} F(U) & \xleftarrow{f(U,\ V)} & G(V) \\ \rho^U_{U'} \downarrow & & \downarrow \rho^V_{V'} \\ F(U') & \xleftarrow[f(U',\ V')]{} & G(V') \end{array}$$

The collection of all ϕ-morphisms $G \to F$ is denoted by $\mathrm{Hom}_\phi(G, F)$.

7.9 Examples. A. If $\phi = id_X : X \to X$, then a ϕ-morphism: $G \to F$ is just a morphism in Presh/X. Hence $\mathrm{Hom}_{id_X}(G, F) = \mathrm{Hom}(G, F)$.

B. If $\phi : X \to pt$ as in 7.5A, then G is just an abelian group, and to give a ϕ-morphism: $G \to F$ is to give an Abgp-morphism: $G \to \Gamma(X, F)$; for then all the other morphisms follow from the commutativity of the triangle

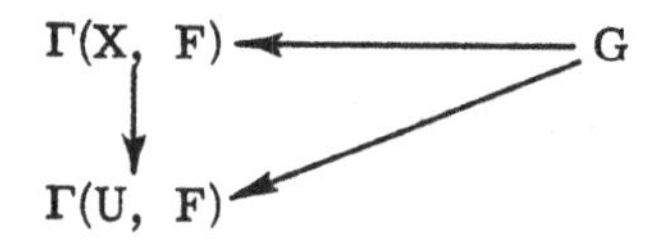

For instance, if G, F were sheaves of rings (cf. 1.6E) and $f : G \to F$ was a ϕ-morphism of sheaves of rings (that is, all the $f(U, V)$ are ring morphisms), then to give f is to give F the structure of a sheaf of G-algebras (cf. 1.6E).

C. For any $\phi : X \to Y$ and presheaf F on X, there is a natural ϕ-morphism $\overline{\phi} : \phi_* F \to F$, given by

$$\overline{\phi}(U,\ V) = \rho_U^{\phi^{-1}V} : (\phi_* F)(V) \to F(U) \qquad \text{for } U \subseteq \phi^{-1}V.$$
$$= F(\phi^{-1}V)$$

7.10 Proposition. *Given a continuous* $\phi : X \to Y$ *and presheaves* F, G *on* X, Y *respectively, there is a natural bijection*

$$\mathrm{Hom}_Y(G,\ \phi_* F) \xrightarrow{\sim} \mathrm{Hom}_\phi(G,\ F)$$

given by 'composition' with the natural ϕ-*morphism* $\overline{\phi} : \phi_* F \to F$.

Proof. A ϕ-morphism $f : G \to F$ is in fact determined by the

$f(\phi^{-1}V, V)$ for V open in Y, since then the $f(U, V)$ for $U \subseteq \phi^{-1}V$ are prescribed by the commutativity of the triangle

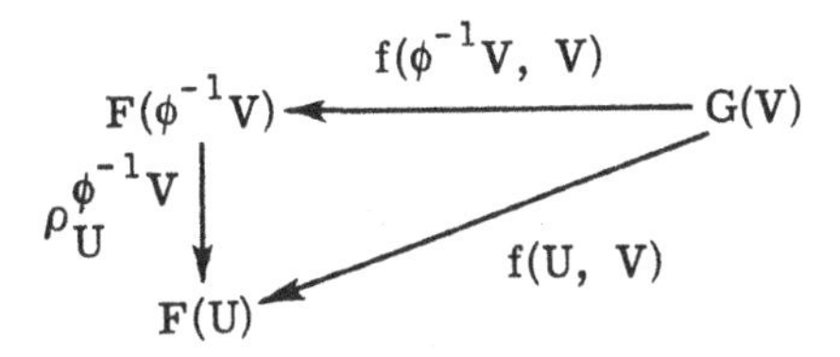

(putting $V' = V$ in square of 7.8). But the $f(\phi^{-1}V, V) : G(V) \to F(\phi^{-1}V) = (\phi_*F)(V)$ just constitute a morphism $G \to \phi_*F$ in Presh/Y. //

7.11 **Construction.** Given a continuous map $\phi : X \to Y$, and a presheaf G on Y, we can construct a sheaf ϕ^*G on X, functorially in G (called the inverse image of G by ϕ), together with natural morphisms:

(i) for any presheaf G on Y, $G \to \phi_*\phi^*G$

(ii) for any sheaf F on X, $\phi^*\phi_*F \to F$

(that is, natural transformations from $id_{Presh/Y}$ to $\phi_*\phi^*$ and from $\phi^*\phi_*$ to $id_{Shv/X}$).

It may help to keep the examples of 7.5 in mind. First construct the sheaf space LG as in 2.3.8 to obtain a diagram of continuous maps and topological spaces:

$$\begin{array}{ccc} & & LG \\ & & \downarrow p \\ X & \xrightarrow{\phi} & Y \end{array}$$

Now let

$$E = \{(e, x) \in LG \times X;\ p(e) = \phi(x)\}$$

with the topology induced from the product topology on $LG \times X$. Thus we have a commutative diagram of continuous maps:

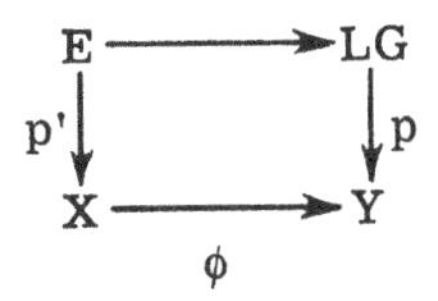

(with $p' : (e, x) \mapsto x$).

I claim that (E, p') is a sheaf space over X; that is, p' is a local homeomorphism, for

$$(e, x) \in E \Rightarrow \exists \text{ open neighbourhood } W \text{ of } e \text{ and inverse homeomorphisms } W \underset{\sigma}{\overset{p|W}{\rightleftarrows}} V \text{ with } V \text{ open in } Y.$$

$$\Rightarrow E \cap W \times \phi^{-1}V \underset{(\phi \circ \sigma, \mathrm{id})}{\overset{p'}{\rightleftarrows}} \phi^{-1}(V) \text{ are inverse homeomorphisms.}$$

Let $\phi^*G = \Gamma E$ be the corresponding sheaf (2. 3. 6); then in fact we can interpret the sections of ϕ^*G: for U open in X,

$$\Gamma(U, \phi^*G) = \{\text{continuous maps } \sigma : U \to E;\ p' \circ \sigma = \mathrm{id}_U\}$$
$$\cong \{\text{continuous maps } \sigma' : U \to LG;\ p \circ \sigma' = \phi|U\},$$

and we see from this that ϕ^*G is an abelian sheaf (2. 5. 2).

7. 12 **Lemma.** The continuous map $E \to LG$ induces isomorphisms

$$(\phi^*G)_x \overset{\sim}{\to} G_{\phi(x)}$$

for $x \in X$.

Proof. This is clear, since $p'^{-1}(x) = p^{-1}(\phi(x)) \times \{x\}$. //

Continuing 7. 11, if $f : G \to G'$ is a morphism of Presh/Y, we get a diagram

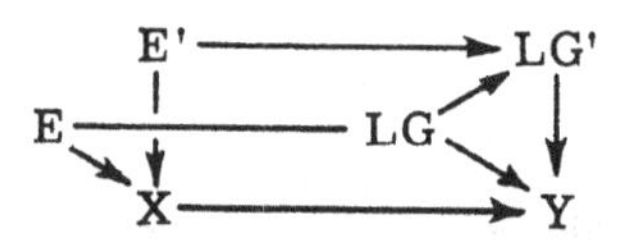

and we obtain easily a sheaf space morphism $E \to E'$ and so a sheaf morphism $\phi^*G \to \phi^*G'$. (Checking $\phi^*(\mathrm{id}) = \mathrm{id}$, $\phi^*(f \circ g) = \phi^*f \circ \phi^*g$ is easy, as usual).

Now for the natural transformation (i): $\mathrm{id}_{\text{Presh}/Y} \to \phi_*\phi^*$. We have a natural ϕ-morphism $G \to \phi^*G$, given over V open in Y by

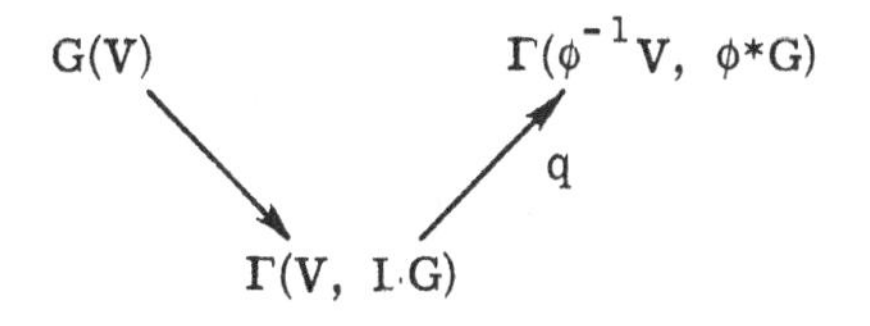

where $q : \sigma \to (\sigma' : x \to (\sigma(\phi(x)), x))$, and σ' is continuous since it is the composite

$$\phi^{-1}V \underset{(\phi,\ \mathrm{incl})}{\to} V \times X \underset{\sigma \times 1}{\to} E(\hookrightarrow LG \times X).$$

Hence by 7.10 there is a natural morphism $G \to \phi_*\phi^*G$.

Now for the natural transformation (ii): $\phi^*\phi_* \to \mathrm{id}_{Shv/X}$. If F is a sheaf on X, and V is open in Y, we have maps

$$\Gamma(V, \phi_*F) = \Gamma(\phi^{-1}V, F) \to F_x$$

if $x \in \phi^{-1}V$. If we fix $x \in X$ and let V run through all open sets of Y with $\phi(x) \in V$, we obtain a target

$$\Gamma(V, \phi_*F) \to F_x$$

for the direct system defining $(\phi_*F)_{\phi(x)}$. Hence there is an abelian group morphism $r_x : (\phi_*F)_{\phi(x)} \to F_x$ (not in general isomorphism; cf. 7.5C). Now apply the ϕ^* construction to ϕ_*F:

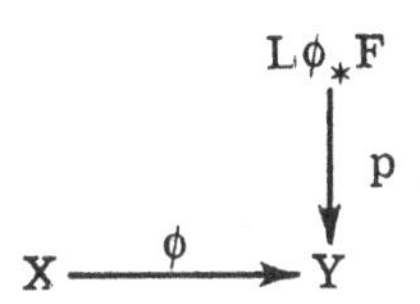

We can map

$$\Gamma(U, \phi^*\phi_*F) \cong \{\text{continuous } \sigma : U \to L\phi_*F;\ p \circ \sigma = \phi\} \to \Gamma(U, LF)$$

by

$$\sigma \mapsto \sigma'$$

where $\sigma'(x) = r_x(\sigma(x))$ for $x \in U$ (so that $\sigma(x) \in (\phi_*F)_{\phi(x)}$).

We must check that σ' is continuous: to obtain $r_x(\sigma(x))$ we pick a neighbourhood V of $\phi(x)$, and some $\chi \in \Gamma(V, \phi_*F)$ such that $\chi_{\phi(x)} = \sigma(x)$; since σ is continuous we may assume, by shrinking V, that $\forall x' \in \phi^{-1}(V)$ $\chi_{\phi(x')} = \sigma(x')$; but then $\sigma'(x) = r_x\sigma(x) = \chi_x$, where we now regard $\chi \in \Gamma(V, \phi_*F)$ as a member of $\Gamma(\phi^{-1}V, F)$. However, χ will suffice to evaluate $\sigma'(x')$ for any $x' \in \phi^{-1}(V)$, so that in $\phi^{-1}(V)$, σ' agrees with $\hat{\chi} : x' \mapsto \chi_{x'}$, and so σ' is continuous at x.

It follows easily that we have defined a natural morphism $\phi^*\phi_*F \to F$. The diagram shows the sheaf spaces involved:

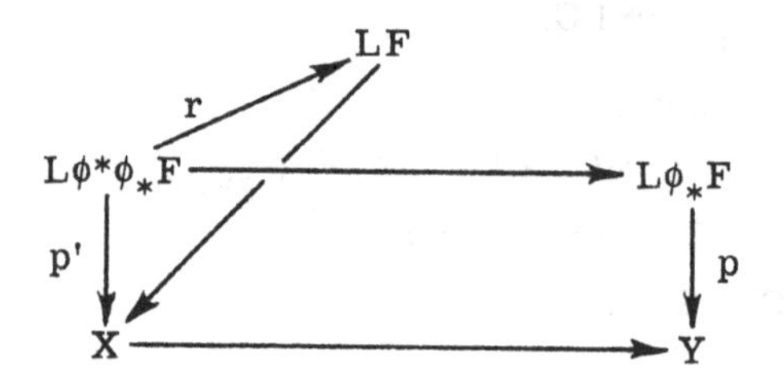

r is the morphism such that $r(e) = r_x\psi_x(e)$ if $x = p'(e)$ where $\psi_x : (\phi^*\phi_*F)_x \to (\phi_*F)_{\phi(x)}$ is the natural morphism of 7.12.

7.13 Theorem. *If* $\phi : X \to Y$ *is a continuous map, and* G *is a presheaf on* Y, *then the sheaf* ϕ^*G *with the morphism* $n : G \to \phi_*\phi^*G$ *has the universal property that for any sheaf* F *on* X, *the map*

$$\mathrm{Hom}_X(\phi^*G, F) \to \mathrm{Hom}_Y(G, \phi_*F) : f \mapsto (\phi_*f) \circ n$$

is bijective.

Furthermore this property characterises ϕ^*G *uniquely up to isomorphism in* Shv/X.

Thus we have natural isomorphisms:

$$\mathrm{Hom}_X(\phi^*G, F) \xrightarrow{\sim} \mathrm{Hom}_\phi(G, F) \xrightarrow{\sim} \mathrm{Hom}_Y(G, \phi_*F)$$

whenever F *is a sheaf on* X *and* G *a presheaf on* Y.

Proof. Given $f : \phi^*G \to F$ we obtain $G \xrightarrow{n} \phi_*\phi^*G \xrightarrow{\phi_*f} \phi_*F$. The inverse map is: given $g : G \to \phi_*F$, compose $\phi^*G \xrightarrow{\phi^*g} \phi^*\phi_*F \to F$ (7.11(ii)). That these procedures are inverse may be seen by considering the procedures:

given $f : \phi^*G \to F$ we get a diagram for V open in Y, and the dotted map gives $G \to \phi_*F$;

$$\begin{array}{ccc} \Gamma(\phi^{-1}V, \phi^*G) & \longleftarrow & G(V) \\ \downarrow & & \vdots \\ \Gamma(\phi^{-1}V, F) & = & \Gamma(V, \phi_*F) \end{array}$$

given $g: G \to \phi_* F$ we get $\forall x \in X$:
and the dotted maps give
$\phi^* G \to F$ (and determine it
uniquely since $\phi^* G$ and F
are both sheaves).

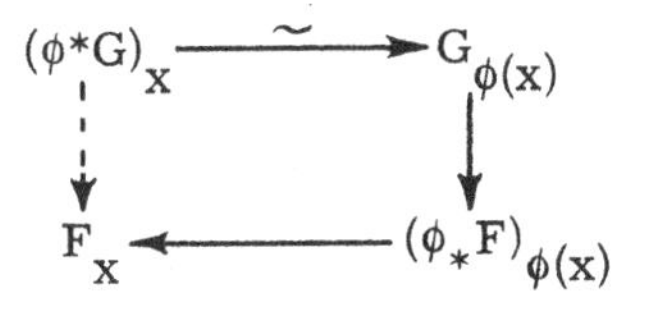

For the uniqueness, if also $\bar{n} : G \to \phi_* \bar{G}$ has the property, then n, $\bar{n}$ factor through each other: say

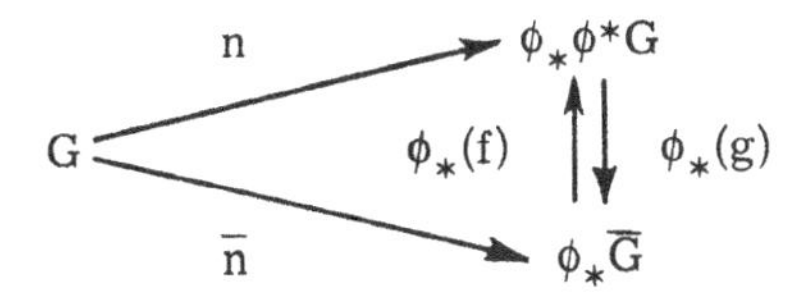

Then $g \circ f : \bar{G} \to \bar{G}$ has $\phi_*(g \circ f) \circ \bar{n} = \phi_*(\mathrm{id}_{\bar{G}}) \circ \bar{n}$ and so $g \circ f = \mathrm{id}_{\bar{G}}$; similarly $f \circ g = \mathrm{id}_{\phi^* G}$, and so $\phi^* G \xrightarrow{\sim} \bar{G}$. //

7.14 **Scholium.** In the language of adjoint functors, 7.13 says that in the diagram $\mathrm{Shv}/X \underset{\phi_*}{\overset{\phi^*}{\rightleftarrows}} \mathrm{Presh}/Y$, ϕ^* is <u>left adjoint</u> to ϕ_*.

7.15 **Remark.** If we had worked harder and proved the existence of $\phi^* G$ satisfying 7.13 independently of Chapter 2 and the L construction, we could have deduced the sheafification of a presheaf G on Y as $(\mathrm{id}_Y)^* G$; for if $\phi = \mathrm{id}_Y$, 7.13 reduces to 2.4.2.

7.16 **Proposition.** <u>The functor</u> $\phi^* : \mathrm{Shv}/Y \to \mathrm{Shv}/X$ <u>is exact.</u>

Proof. An exact sequence of sheaves on Y $0 \to F \to G \to H \to 0$ gives by 7.12 for each $x \in X$ a commutative diagram:

$$\begin{array}{ccccccccc} 0 & \to & F_{\phi(x)} & \to & G_{\phi(x)} & \to & H_{\phi(x)} & \to & 0 \\ & & \uparrow & & \uparrow & & \uparrow & & \\ 0 & \to & (\phi^* F)_x & \to & (\phi^* G)_x & \to & (\phi^* H)_x & \to & 0 \end{array}$$

with the vertical maps all isomorphisms. Since the top line is exact, so is the bottom line, and so $0 \to \phi^* F \to \phi^* G \to \phi^* H \to 0$ is an exact sequence of sheaves by 6.5(ii). //

7.17 **Remark.** After 7.14, we should know from the properties of adjoint functors that ϕ_* is left exact and ϕ^* is right exact. The fact that ϕ^* is also left exact is special to this situation.

3.8 Restriction and extension

8.1 **Definition.** If X is a subspace of Y, with inclusion map $\phi : X \hookrightarrow Y$, and G is a sheaf on Y, the sheaf ϕ^*G is called the *restriction* of G to X and is denoted by $G|X$. If X is open in Y, this coincides with the notion of 2. Ex. 3.

8.2 **Proposition.** *If* $\phi : X \hookrightarrow Y$ *and* F *is a sheaf on* X, *then the morphism* $(\phi_*F)|X \to F$ *of 7.11(ii) is an isomorphism in* Shv/X.

Proof. The morphism is given on stalks by the $r_x : (\phi_*F)_x \to F_x$ (for $x \in X$) which are defined by the target of the direct system obtained as V runs through all open sets of Y such that

$$\begin{array}{ccc} \Gamma(V, \phi_*F) & \longrightarrow & F_x \\ \| & \nearrow & \\ \Gamma(\phi^{-1}V, F) & & \end{array}$$

$V \ni x$. Hence $\phi^{-1}V = V \cap X$ runs through all open neighbourhoods of x in X, and so this is the direct system defining F_x. Hence each r_x is an isomorphism, and the result follows from 4.10. //

8.3 **Theorem.** *If* $\phi : X \hookrightarrow Y$ *and* G *is a sheaf on* Y *with sheaf space* (LG, p), *then* $G|X$ *is a sheaf with sheaf space* $(p^{-1}X, p|p^{-1}X)$. $G|X$ *has the universal property that for any sheaf* F *on* X, *any morphism* $g : G|X \to F$ *extends to a unique morphism* $f : G \to \phi_*F$ *whose restriction* ϕ^*f *to* X *is the morphism* g.

Proof. The first part follows directly from the construction of ϕ^*G (7.11). The second part is a restatement of 7.13 in this situation, taking account of 8.2. //

8.4 **Definition.** If $X \subseteq Y$ and F is a sheaf on X, a sheaf F' on Y is called an *extension* of F to Y iff $F'|X \cong F$; so 8.2 shows that ϕ_*F is an extension of F. F' is called an *extension of* F *by zero* iff $F'|X \cong F$ and $\forall y \in Y\backslash X$ $F'_y = 0$ (so that $F'|Y\backslash X = 0$ by 8.3). Example 7.5B shows that ϕ_*F need not be an extension of F by zero.

8.5 **Definition.** A subspace X of Y is called locally closed iff the following equivalent conditions hold:

(a) $\forall x \in X \ \exists U \ni x$, U open in Y such that $U \cap X$ is closed in U;

(b) $\exists U$ open in Y and V closed in Y such that $X = U \cap V$;

(c) X is open in its closure $\overline{X}$.

For example, $\mathbf{R}\setminus\{0\} \hookrightarrow \mathbf{R} \hookrightarrow \mathbf{R}^2$ is locally closed, but $\mathbf{Q} \hookrightarrow \mathbf{R}$ is not. If X is locally closed in Y, and Y is locally closed in Z, then X is locally closed in Z.

8.6 **Theorem.** $X \subseteq Y$ is locally closed iff it has the property:

(*86) for any sheaf F on X, there is (up to isomorphism) a unique extension of F by zero, denoted F^Y. (Some authors denote it by $j_!F$ where $j : X \hookrightarrow Y$.)

Proof. Suppose X has the property (*86). Then in particular, putting $F = Z_X$ the constant sheaf on X, there is a sheaf G on Y with stalks

$$G_y = \begin{cases} Z & \text{if } y \in X \\ 0 & \text{if } y \in Y\setminus X. \end{cases}$$

Given $x \in X$ we can find an open U in Y and a section $s \in \Gamma(U, G)$ such that $s(x) = 1$; by shrinking U we can ensure that $\forall x' \in U \cap X$ $s(x') = 1$ (since $G|X \cong Z_X$). But then $U \cap (Y\setminus X) = \{y \in U;\ s(y) = 0\}$, and this is open in U (for if $s(y) = 0$ then $s(y') = 0$ for y' in some neighbourhood of y). Hence X is locally closed.

Now suppose that X is locally closed in Y, and that F' is an extension of F by zero. Then for U open in Y we have an injective map: $\Gamma(U, F') \to \Gamma(U \cap X, F) : s' \mapsto s'|U \cap X$ and a section $s \in \Gamma(U \cap X, F)$ is in the image of this map iff s remains continuous when it is extended to U by the definition: $s(y) = 0$ if $y \in U\setminus(U \cap X)$. But this will hold iff $\{x \in U \cap X;\ s(x) \neq 0\}$ is closed in U, a condition which is independent of the particular F' chosen. Hence if F^Y exists, it is unique up to isomorphism.

We shall show that F can be extended to Y by first extending F to $\overline{X}$, and thence to Y. So assume now that X is open in Y. Let (LF, p) be the sheaf space of F over X, and set

$$E = LF \amalg Y/\sim ,$$

where $\sim$ is the smallest equivalence relation with $x \sim 0_x$ for all $x \in X$, where 0_x is the zero element of the stalk $p^{-1}(x)$ at x. Give E the quotient topology and let $p' : E \to Y$ be the natural map. Since X is open in Y, p' is easily seen to be a local homeomorphism. Letting $F^Y = \Gamma E$ be the associated sheaf, we see that F^Y is an extension of F by zero.

The proof is completed by the following Lemma, which enables us to extend $F^{\overline{X}}$ from $\overline{X}$ to Y if X is any locally closed subspace of Y. //

8.7 **Lemma.** If $\phi : X \hookrightarrow Y$ and X is closed in Y, then $\phi_* F$ is the extension of F by zero.

Proof. By 8.2 $\phi_* F | X \cong F$. If $y \in Y \backslash X$, then since for any open U with $y \in U \subseteq Y \backslash X$ we have

$$\Gamma(U, \phi_* F) = \Gamma(\phi^{-1} U, F) = \Gamma(U \cap X, F) = \Gamma(\emptyset, F) = 0$$

we see that $(\phi_* F)_y = 0$ as required. //

8.8 **Corollary.** Let X be a subspace of Y and F a sheaf on X.

(a) If X is closed in Y, then for U open in Y

$$\Gamma(U, F^Y) \xrightarrow{\sim} \Gamma(U \cap X, F) : \sigma \mapsto \sigma | U \cap X$$

is an isomorphism.

(b) If X is open in Y, then LF is an open subspace of LF^Y (namely $p^{-1}X$, if $p : LF^Y \to Y$ is the structure map).

Proof. (a) By 8.7 $F^Y \cong \phi_* F$ and $\Gamma(U, \phi_* F) = \Gamma(U \cap X, F)$ by definition of ϕ_*. Aliter: by the proof of 8.6 the image of the map contains $s \in \Gamma(U \cap X, F)$ iff $\{x \in U \cap X;\ s(x) \neq 0\}$ is closed in U; but this set is always closed in $U \cap X$, and so if X is closed it is also closed in U.

(b) This follows from the topology placed on $LF^Y = E$ in the proof of 8.6. //

8.9 **Proposition.** *If X is a locally closed subspace of Y, the functor*

$$Shv/X \to Shv/Y : F \mapsto F^Y$$

is exact.

Proof. The functoriality of $F \mapsto F^Y$ follows easily from the constructions of 8.6 and 8.7. An exact sequence of sheaves $0 \to F \to G \to H \to 0$ over X gives a sequence

$$0 \to F^Y \to G^Y \to H^Y \to 0 \qquad (*)$$

and the stalk sequence of this over $y \in Y$ is

$$0 \to F_y \to G_y \to H_y \to 0 \quad \text{if } y \in X$$

and

$$0 \to 0 \to 0 \to 0 \to 0 \quad \text{if } y \notin X$$

each of which is exact, so (*) is exact in Shv/Y. //

8.10 **Corollary.** *If X is a locally closed subspace of Y, the functor*

$$Shv/Y \to Shv/Y : G \mapsto (G|X)^Y$$

is exact (some writers denote $(G|X)^Y$ *by* G_Y).

Proof. It is the composite of the functors $-^Y$ and ϕ^* where $\phi : X \hookrightarrow Y$, each of which is exact (by 8.9 and 7.16). //

8.11 **Theorem.** *If U is an open subspace of X and F is a sheaf on X, there is a short exact sequence in* Shv/X:

$$0 \to (F|U)^X \to F \to (F|X\backslash U)^X \to 0.$$

Proof. That there is a morphism $(F|U)^X \to F$ follows from the fact that the inclusion $L(F|U)^X \hookrightarrow LF$ is open and so a morphism of sheaf spaces (2.3.5(c)); for

$$L(F|U)^X = \{0_x \in F_x; x \in X\} \cup p^{-1}(U)$$

where $p : LF \to X$ is the projection and by 8.8(b) $p^{-1}U$ has the subspace topology inside LF.

Putting $C = X \backslash U$, by 8.8(a) there are maps (for V open in X)

$$\Gamma(V, F) \to \Gamma(V \cap C, F|C) \cong \Gamma(V, (F|C)^X)$$

given by $\sigma \mapsto \sigma|V \cap C$; hence there is a sheaf morphism $F \to (F|C)^X$.

The exactness of the sequence follows from the fact that the stalk sequence for $x \in X$ is

$$0 \to F_x \xrightarrow{id} F_x \to 0 \to 0 \qquad \text{if } x \in U$$

and

$$0 \to 0 \to F_x \xrightarrow{id} F_x \to 0 \qquad \text{if } x \in X\backslash U$$

and each of these is exact. //

8.12 **Proposition.** Suppose that X is an open subspace of Y and F is a sheaf on X. Then for each extension F' of F to Y there are unique morphisms g, h such that the diagram

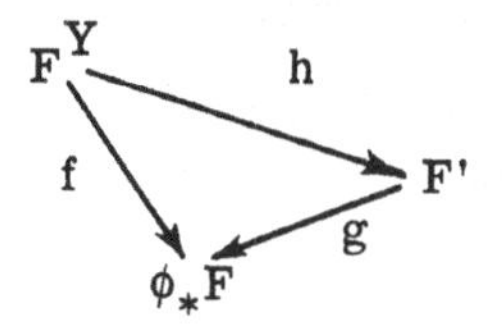

commutes, and gives on restricting to X a commutative diagram

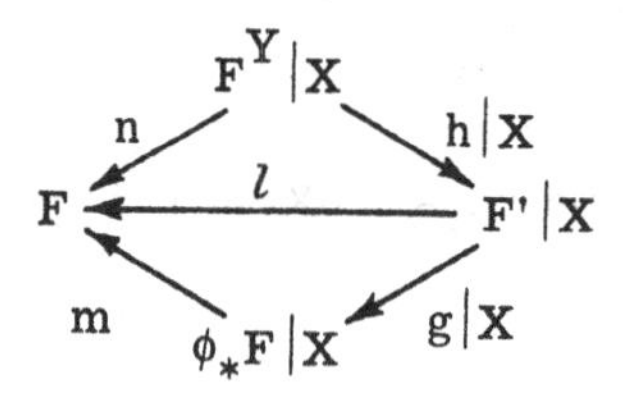

where n is the isomorphism of 8.6,

m is the isomorphism of 8.2

and l is the isomorphism which shows that $F'|X \cong F$.

Proof. By the universal property of ϕ_* (8.3 or 7.13), the morphisms $F'|X \xrightarrow{l} F$ and $F^Y|X \xrightarrow{n} F$ give rise to unique morphisms $F' \xrightarrow{g} \phi_*F$ and $F^Y \xrightarrow{f} \phi_*F$ such that the triangles

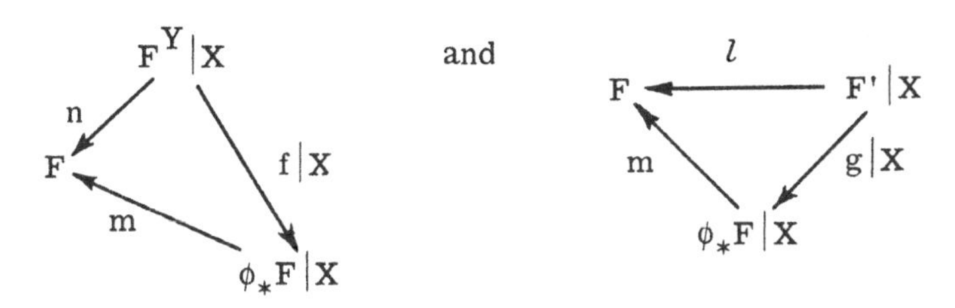

commute. For U open in Y we have maps

$$\Gamma(U, F^Y) \to \Gamma(U, F')$$
$$\sigma \mapsto \sigma' \quad \text{where} \quad \sigma'(y) = \begin{cases} \sigma(y) & \text{if } y \in X \\ 0 & \text{if } y \notin X \end{cases}$$

for $y \in Y$ (using the identification l); σ' is continuous since if $y \in X$, σ' agrees on the open neighbourhood X of y with σ, and hence σ' is continuous at y, while if $y \in Y\backslash X$ and $\sigma'(y) = \tau(y)$ with $\tau \in \Gamma(U, F')$ then τ is zero on some neighbourhood of y, and so is σ' since σ is continuous, so that σ' is continuous at y. Hence there is a sheaf morphism $F^Y \xrightarrow{h} F'$ such that the diagram

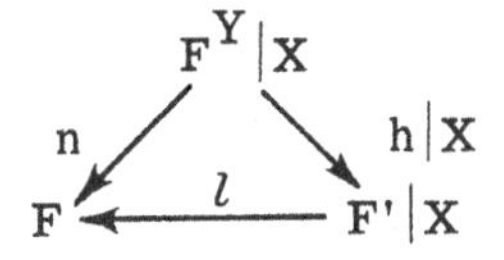

commutes. The result follows. //

8.13 **Remark.** In categorical terms, 8.12 shows that if X is open in Y and $\phi : X \hookrightarrow Y$, the category of extensions of a sheaf F on X to Y has a final object ϕ_*F (this is in fact true for any subspace X of Y, as the proof shows), and an initial object F^Y. This in fact shows that for open subspaces as well as having a right adjoint ϕ_*, the functor

$\phi^* = -|X : Shv/Y \to Shv/X$ has a left adjoint $-^Y$, (which is exact, by 8.9).

8.14 **Example.** Let $X = \mathbf{R}\setminus\{0\} \overset{\phi}{\hookrightarrow} \mathbf{R} = Y$ and let G be a constant sheaf on Y with typical fibre an abelian group A. Then $F = G|X$ is the constant sheaf with fibres A over X, and F^Y has stalks

$$(F^Y)_y = \begin{cases} A & \text{if } y \neq 0 \\ 0 & \text{if } y = 0 \end{cases}$$

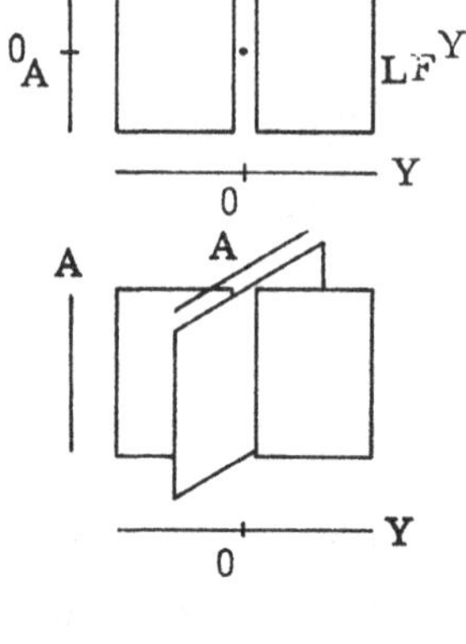

while $\phi_* F$ has stalks

$$(\phi_* F)_y = \begin{cases} A & \text{if } y \neq 0 \\ A \oplus A & \text{if } y = 0 \end{cases}$$

(since if U is a small open interval around 0, $U \cap X$ has two components).

Exercises on Chapter 3

1. Let P be the category of pointed sets, whose objects are the pairs (A, a) with $a \in A \in \mathrm{Ob(Sets)}$, and whose morphisms $(A, a) \to (B, b)$ are the maps of sets $f : A \to B$ such that $f(a) = b$. Show that P is a category with a zero object, kernels and cokernels, but in which not every epimorphism is a cokernel.

2. For X any topological space, show that the following presheaves of sets over X are in fact sheaves:

(a) Fixing an open $V \subseteq X$, let

$$h_V(U) = \begin{cases} \text{singleton} & \text{if } U \subseteq V \\ \emptyset & \text{if not} \end{cases} \qquad \text{for } U \text{ open in } X,$$

with the unique restrictions.

(b) Let $\Omega(U) = \{W;\ W \text{ open in } X \text{ and } W \subseteq U\}$ for U open in X with restriction: $\Omega(U) \to \Omega(V) : W \mapsto W \cap V$.

Interpreting presheaves on X as contravariant set-valued functors on $\mathfrak{U}$, the category of open sets of X (1.6E), show that the h_V defined by (a) are the <u>representable</u> functors

$$h_V = \mathrm{Hom}_{\mathcal{U}}(-, V) : U \mapsto \mathrm{Hom}_{\mathcal{U}}(U, V).$$

Interpret the Yoneda Lemma ([Macl] III. 2, [Mit] IV. 2) as saying that

$$\mathrm{Hom}_{P/X}(h_V, F) \xrightarrow{\sim} F(V)$$

for any presheaf F on X, where P/X denotes the category of set-valued presheaves on X (that is, the functor category $\mathrm{Sets}^{\mathcal{U}^{op}}$, where $\mathcal{U}^{op}$ is the dual of $\mathcal{U}$). Putting $F = h_U$, this shows that $V \mapsto h_V$ is a full and faithful embedding of $\mathcal{U}$ into P/X.

3. For a topological space X, let P/X and S/X be the categories of presheaves and sheaves of sets over X. Imitate the results of Chapter 3 as follows:

(a) Show that $F \to G$ is a monomorphism in P/X (or in S/X) iff for all open U $F(U) \to G(U)$ is injective iff for all $x \in X$ $F_x \to G_x$ is injective.

(b) Show that $F \to G$ is an epimorphism in P/X iff for all open U $F(U) \to G(U)$ is surjective; while $F \to G$ is an epimorphism in S/X iff $\forall x \in X$ $F_x \to G_x$ is surjective.

(c) Discover an appropriate definition of 'equivalence relation' $\sim$ on a presheaf F so that you can define a presheaf $F/\sim$ (respectively a sheaf $F/\sim$ if F is a sheaf) together with an epimorphism $F \to F/\sim$ (in the appropriate category) with a universal property.

[This is an analogue of cokernel; if necessary, see [G] Ch. II, 1. 9.]

(d) Show that each of the categories P/X and S/X has all limits and all colimits (in particular all equalisers and coequalisers); that in each category every monomorphism is an equaliser and every epimorphism is a coequaliser; and that every morphism has a unique epimorphism-monomorphism factorisation.

[For definitions, see [Macl] III, 3 and 4; V. 2. These results are the analogue of the abelianness of Presh/X and Shv/X. See also Q5 below.]

(e) Show that for a continuous map $\phi : X \to Y$ there are induced functors $S/X \underset{\phi^*}{\overset{\phi_*}{\rightleftarrows}} S/Y$ (as in §3. 7) with the adjointness property 7. 13, and such that ϕ^* preserves all finite limits (analogue of 'ϕ^* is left exact').

(f) Is there an analogue of extension by zero?

4. Let F, G be presheaves of sets on a topological space X. Define a new presheaf $\underline{\mathrm{Hom}}(F, G)$ with

$$\underline{\mathrm{Hom}}(F, G)(U) = \mathrm{Hom}_{\mathrm{Sets}}(F(U), G(U))$$

for U open in X. Show that if F and G are both sheaves, then so is $\underline{\mathrm{Hom}}(F, G)$.

Prove that this construction has the following universal property: for F, G, H presheaves (respectively sheaves) of sets on X, there is a bijection

$$(*4) \qquad \mathrm{Hom}(H, \underline{\mathrm{Hom}}(F, G)) \cong \mathrm{Hom}(H \times F, G)$$

natural in F, G and H. Here $H \times F$ is the product object provided by Q3(d); it is constructed 'pointwise'. [This shows that for each presheaf (respectively sheaf) F, the functor $\underline{\mathrm{Hom}}(F, -)$ is right adjoint to $- \times F$; compare the situation in Sets, where the same is true for the functor

$$\mathrm{Hom}(F, -) : G \mapsto G^F = \text{set of maps} : F \to G.]$$

Show that, if we want the property (*4) to hold, then the definition of $\underline{\mathrm{Hom}}$ is forced on us, for presheaves at least. [Hint: put $H = h_U$ and use the Yoneda lemma (Q2).]

Reinterpret (*4) as requiring the existence of an <u>evaluation map</u>

$$\underline{\mathrm{Hom}}(F, G) \times F \to G$$

with a suitable universal property.

5. Show that each of the categories P/X, S/X of Q3 has a <u>subobject classifier</u>; that is, an object Ω with the equivalent properties (prove their equivalence):

(a) there is a natural bijection

$$\mathrm{Hom}(F, \Omega) \cong \text{set of subobjects of } F$$

for any object F;

(b) Ω has a special subobject $1 \xrightarrow{t} \Omega$ such that any monomorphism $G \to F$ is the pullback of t over a unique morphism $F \to \Omega$ (called the classifying map of $G \to F$).

[Hint: for P/X, put $F = h_U$ and use the Yoneda lemma (Q2) to discover Ω; for S/X use the Ω of Q2(b).]

Compare the situation in Sets $= S/X$ for X a one-point space, where Ω is a two-point set and the classifying map is the characteristic map of a subset.

[Q3(d), Q4 and Q5 show that P/X and S/X each satisfy the axioms of an elementary topos; see Kock and Wraith: Elementary Toposes (Aarhus Lecture Notes No. 30, 1971); Freyd: Aspects of Topoi (Bull. Aust. Math. Soc., 7 (1972) 1-76).]

6. Since any category of presheaves is a functor category, we can define it for categories other than those of open sets of a topological space. For C any category, let P/C be the category $\text{Sets}^{C^{op}}$ of contravariant functors $C \to$ Sets (and natural transformations) (here C^{op} denotes the dual of C, which has the same objects and morphisms, but has composition defined in the reverse order to that in C).

As a special case, any group G can be regarded as a category with one object whose endomorphisms are the elements of G, with composition defined as in G (hence every morphism is an isomorphism). Show that the category P/G can be regarded as the category of sets-with-a-G^{op}-action, that is the category of permutation representations of G.

7. If A and B are abelian categories, a functor $F : A \to B$ is called half exact iff whenever $0 \to P \to Q \to R \to 0$ is exact in A, $FP \to FQ \to FR$ is exact in B. F is called additive iff all the maps $F : \text{Hom}_A(P, Q) \to \text{Hom}_B(FP, FQ)$ are abelian group homomorphisms. Consider the conditions:

(a) F is half exact

(b) F preserves biproducts (that is $F(P \oplus Q) \cong FP \oplus FQ$)

(c) F is additive.

Show that (a) $\Rightarrow$ (b) $\Longleftrightarrow$ (c).

8. Let K be an abelian category and X a topological space. Let K-Presh/X denote the category of K-valued presheaves on X (1. 6E). Show that it is an abelian category.

Show that we can define the full subcategory K-Shv/X of K-valued sheaves on X by the following condition: a presheaf F is a sheaf iff for each object T of K the presheaf of sets

$$U \mapsto \mathrm{Hom}(T, F(U)) \qquad \text{(for } U \text{ open in } X\text{)}$$

is a sheaf of sets. (Verify that if K has the appropriate products, this condition coincides with that of 2. 1. 4 or 2. 1. 8.)

What extra conditions do you need on K to ensure that K-Shv/X is an abelian category? To ensure that the statement of 2. 5 holds for K in place of Abgp? (Consider, for example, K= the category of all finite abelian groups.)

9. Let $0 \to P \to Q \to R \to 0$ be an exact sequence of presheaves of abelian groups over a topological space. Show that

(a) if Q is a sheaf and R is a monopresheaf, then P is a sheaf;

(b) if P is a sheaf and Q is a monopresheaf, then R is a monopresheaf.

10. A sheaf F of abelian groups on X is called locally free if and only if each point $x \in X$ has an open neighbourhood U in X such that the sheaf $F|U$ is isomorphic to a constant sheaf with typical stalk a free abelian group (of finite rank). Show that if X is connected, a locally free sheaf has a well-defined rank. Show by example that a locally free sheaf on X need not be isomorphic to a constant sheaf even if X is connected, and that ϕ_* does not preserve the property of being locally free. Prove however that the inverse image of a locally free sheaf of rank n is locally free and of the same rank.

[Recall that a group G is free abelian if and only if there is a natural number n (called the rank of G) such that G is isomorphic to $\mathbf{Z}^n = \oplus_{i=1}^n \mathbf{Z}$. You may assume that 'rank' is well-defined for such groups.]

4 · Ringed spaces

This chapter brings us to the essential core of geometry, as expressed in the language of sheaf theory. We study spaces equipped with a sheaf of rings, and particularly the geometric spaces, where the stalks are all local rings: we show that there is some justification for this name, since morphisms between geometric spaces specialise to the appropriate kinds of maps between several types of manifolds (differentiable, analytic, and so on).

We construct a universal geometric space associated with each commutative ring, and this leads us to the definition of schemes, which are central in modern algebraic geometry.

We then consider sheaves of modules over a ringed space, which generalise the idea of vector bundles, and globalise the idea of a module over a ring. The module constructions of direct sum and product, tensor product and module of homomorphisms also globalise to these sheaves, with appropriate universal properties. Similarly, change of base space by a morphism of ringed spaces gives rise to direct and inverse image functors. Finally, we define the picard group of a ringed space; we shall see later that this can be interpreted as a cohomology group.

Throughout this Chapter the word 'ring' will mean commutative ring with a one, and ring morphisms are required to preserve ones.

4.1 The category of ringed spaces over a ring R

1.1 Recall that if R is a ring, an R-algebra is a ring S together with a ring morphism $\alpha : R \to S$ called the structure map. A morphism of R-algebras $\beta : S \to S'$ is a ring morphism making the triangle

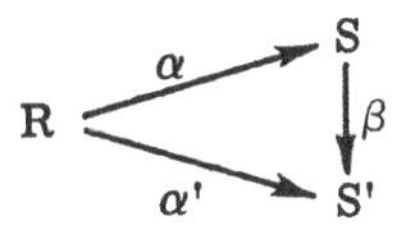

commute. Hence we have a category R-alg of all R-algebras.

For instance, for any ring S there is a unique ring morphism $\mathbf{Z} \to S$ (taking $n \mapsto n1_S$) and any ring morphism $S \to S'$ makes the triangle

commute. Hence any ring is a $\mathbf{Z}$-algebra in a unique way, and indeed the category $\mathbf{Z}$-alg is isomorphic to the category of rings. If R is a field, such as $\mathbf{R}$, $\mathbf{C}$ or the p-adic numbers $\mathbf{Q}_p$ for a prime number $p \in \mathbf{N}$, then an R-algebra is an R-vectorspace together with an R-linear multiplication.

1.2 Definition. If R is a ring, a ringed space over R is a pair $(X, \mathcal{O}_X)$ where X is a topological space and $\mathcal{O}_X$ is a sheaf of R-algebras on X.

Of course a sheaf of R-algebras is a presheaf of R-algebras (defined as in 1.1.2, or 3.1.6E with $K = $R-alg) which satisfies the sheaf conditions (M) and (G) of 2.1.1 and 2.1.3; a morphism of sheaves of R-algebras is defined in the obvious way, following 2.1.13. We can regard $\mathcal{O}_X$ as a sheaf of abelian groups (forgetting the multiplicative structure) and apply the ideas and constructions of Chapters 2 and 3.

By abuse of language, we often say 'X is a ringed space (over R)' and call $\mathcal{O}_X$ the structure sheaf. If no mention is made of R, it is often assumed that $R = \mathbf{Z}$, so that after 1.1 $\mathcal{O}_X$ is just a sheaf of rings.

It is easy to see that the direct limit of a direct system of R-algebras and morphisms has a natural R-algebra structure; in particular the stalks $\mathcal{O}_{X,x}$ for $x \in X$ are R-algebras.

1.3 Proposition. *Suppose that (E, p) is a sheaf space over X such that each fibre $p^{-1}(x)$ $(x \in X)$ has a given R-algebra structure. Then E is the sheaf space of a sheaf of R-algebras iff equivalently*

(a) *for each open U in X, $\Gamma(U, E)$ is an R-algebra under pointwise addition and multiplication of functions;*

(b) *let $E \,\pi\, E = \{(e, e') \in E \times E;\ p(e) = p(e')\}$; then the maps $(e, e') \mapsto e + e'$ and $(e, e') \mapsto ee'$ (ring operations in $p^{-1}(p(e))$) are* continuous: $E \pi E \to E$, and for every $r \in R$ the map $E \to E : e \mapsto re$ is continuous.

Proof. Similar to 2.5.2. //

1.4 **Corollary.** If X and Y are ringed spaces over a ring R, and $\phi : X \to Y$ is a continuous map, then the sheaves $\phi^*\mathcal{O}_Y$ and $\phi_*\mathcal{O}_X$ are sheaves of R-algebras (on X and Y respectively), and the natural morphisms $\mathcal{O}_Y \to \phi_*\phi^*\mathcal{O}_Y$, $\phi^*\phi_*\mathcal{O}_X \to \mathcal{O}_X$ are morphisms of sheaves of R-algebras.

Proof. Easy verification from 3.7.1 and 3.7.11, using 1.3 where necessary. //

1.5 **Examples.**

A. For any topological space X, letting the structure sheaf $\mathcal{O}_X = \mathbf{Z}_X$, the constant sheaf (2.4.6), makes X into a ringed space over $\mathbf{Z}$.

B. Let R be a topological ring, that is a ring endowed with a topology such that the ring operations (of subtraction and multiplication) are continuous; for example $\mathbf{R}$, $\mathbf{C}$, the p-adic numbers $\mathbf{Z}_p$, or any ring with the discrete topology. For any topological space X, let $\mathcal{O}_X = C^R$ be the sheaf of continuous functions on X with values in R (1.2.B and 2.2.B). Then $(X, \mathcal{O}_X)$ is a ringed space over R; the R-algebra structures $R \to \Gamma(U, \mathcal{O}_X)$ for U open in X are given by the constant functions.

For $R = \mathbf{Z}$ with the discrete topology, this gives example A.

C. Let X be a banach space over $\mathbf{R}$ (for instance $\mathbf{R}^n$ for $n \in \mathbf{N}$) and let $\mathcal{O}_X = C^r$ be the sheaf of r-times continuously differentiable $\mathbf{R}$-valued functions on X (cf. 1.2.C, 2.2.B). Then $(X, \mathcal{O}_X)$ is a ringed space over $\mathbf{R}$. Note that for U open in X, the sheaf C^r defined on U as in loc. cit. is just $\mathcal{O}_X|U$ (3.8.1 and 1.4).

References for differentiability: Bourbaki, Variétés différentielles et analytiques, §2; [L] Chapter 1.

D. Let K be a complete non-discrete valued field, such as $\mathbf{R}$, $\mathbf{C}$ or the p-adic numbers $\mathbf{Q}_p$; let X be a banach space over K. Then the structure sheaf $\mathcal{O}_X = C^\omega$ of all K-valued analytic functions on X (cf. 1.2.D and 2.2.B) makes X into a ringed space over K.

References for analytic functions: Bourbaki, Variétés différentielles et analytiques, §§3, 4; Serre, Lie algebras and Lie groups, Chapter LG2.

1.6 **Definition.** Let R be a ring. A <u>morphism</u> $\Phi : (X, \mathcal{O}_X) \to (Y, \mathcal{O}_Y)$ of ringed spaces over R is given by a continuous map $\phi : X \to Y$ of the underlying topological spaces together with a ϕ-morphism of sheaves of R-algebras $\mathcal{O}_Y \to \mathcal{O}_X$; as in 3.7.10 this may be regarded as a morphism $\mathcal{O}_Y \to \phi_*\mathcal{O}_X$ of sheaves of R-algebras on Y.

Given morphisms $(X, \mathcal{O}_X) \to (Y, \mathcal{O}_Y) \to (Z, \mathcal{O}_Z)$ of ringed spaces over R, we get a composite morphism $(X, \mathcal{O}_X) \to (Z, \mathcal{O}_Z)$ by composing the underlying continuous maps $X \xrightarrow{\phi} Y \xrightarrow{\psi} Z$ and combining $\mathcal{O}_Y \to \phi_*\mathcal{O}_X$ and $\mathcal{O}_Z \to \psi_*\mathcal{O}_Y$ to get $\mathcal{O}_Z \to \psi_*\mathcal{O}_Y \to \psi_*\phi_*\mathcal{O}_X = (\psi \circ \phi)_*\mathcal{O}_X$.

Hence we have defined a category of all ringed spaces over R.

1.7 **Proposition.** $\Phi : (X, \mathcal{O}_X) \to (Y, \mathcal{O}_Y)$ <u>is an isomorphism of ringed spaces over</u> R <u>iff</u> Φ <u>is invertible (that is</u> $\exists \Psi$ <u>such that</u> $\Phi \circ \Psi = \mathrm{id}$, $\Psi \circ \Phi = \mathrm{id}$) <u>iff the underlying map</u> $\phi : X \to Y$ <u>is a homeomorphism and the map</u> $\mathcal{O}_Y \to \phi_*\mathcal{O}_X$ <u>is an isomorphism of sheaves of</u> R-<u>algebras.</u>

Proof. The first equivalence is by definition, and the second follows easily. //

1.8 **Example.** Let $(X, \mathcal{O}_X)$ and $(Y, \mathcal{O}_Y)$ be ringed spaces of the kind described in 1.5B (resp. 1.5C, 1.5D), and let $\phi : X \to Y$ be a continuous (resp. differentiable, resp. analytic) map. Then there is a morphism of ringed spaces $(X, \mathcal{O}_X) \to (Y, \mathcal{O}_Y)$ with underlying continuous map ϕ, given by 'composition with ϕ': the ϕ-morphism $\mathcal{O}_Y \to \mathcal{O}_X$ is defined by

$$\Gamma(V, \mathcal{O}_Y) \to \Gamma(\phi^{-1}V, \mathcal{O}_X) : s \mapsto s \circ \phi$$

for V open in Y.

We shall see a qualified converse to this observation in §3.

1.9 **Proposition.** Any morphism $\Phi : (X, \mathcal{O}_X) \to (Y, \mathcal{O}_Y)$ of ringed spaces over R can be factored uniquely as

$$(X, \mathcal{O}_X) \xrightarrow{\Psi} (X, \phi^*\mathcal{O}_Y) \xrightarrow{\Theta} (Y, \mathcal{O}_Y)$$

with Ψ having underlying continuous map id_X and Θ having as morphism of sheaves of R-algebras the natural morphism $\mathcal{O}_Y \to \phi_*\phi^*\mathcal{O}_Y$ (of 3.7.11 and 3.7.13).

Moreover this factorisation has the universal property that if $(X, \mathcal{O}'_X)$ is another ringed space over R through which Φ factors as

$$(X, \mathcal{O}_X) \xrightarrow{\Psi'} (X, \mathcal{O}'_X) \xrightarrow{\Theta'} (Y, \mathcal{O}_Y)$$

with Ψ' having id_X as underlying continuous map, then there is a unique morphism $\phi^*\mathcal{O}_Y \to \mathcal{O}'_X$ making the triangle

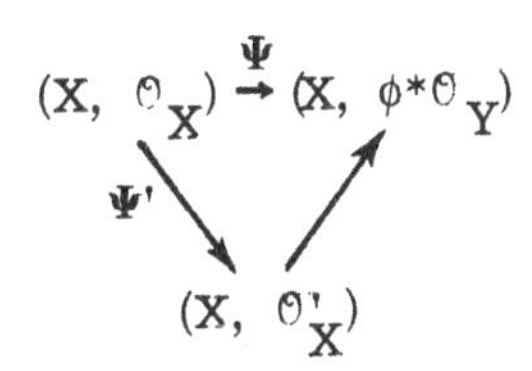

commute.

Proof. The existence and uniqueness of Ψ follow from 3.7.13: the given morphism $\mathcal{O}_Y \to \phi_*\mathcal{O}_X$ associated with Φ corresponds uniquely to a morphism $\phi^*\mathcal{O}_Y \to \mathcal{O}_X$. The universal property of Ψ also follows from 3.7.13, using the bijection

$$\mathrm{Hom}(\mathcal{O}_Y, \phi_*\mathcal{O}'_X) \xrightarrow{\sim} \mathrm{Hom}(\phi^*\mathcal{O}_Y, \mathcal{O}'_X).$$

In each use of 3.7.13 we must make the easy observation that the bijection restricts to a bijection between sets of morphisms of sheaves of R-algebras. //

1.10 **Proposition.** Any morphism $\Phi : (X, \mathcal{O}_X) \to (Y, \mathcal{O}_Y)$ of ringed spaces over R can be factored uniquely as

$$(X, \mathcal{O}_X) \xrightarrow{\Psi} (Y, \mathcal{O}'_Y) \xrightarrow{\Theta} (Y, \mathcal{O}_Y)$$

with Θ having as underlying continuous map id_Y and morphism of sheaves $\mathcal{O}_Y \to id_* \mathcal{O}'_Y = \mathcal{O}'_Y$ a sheaf epimorphism, while the sheaf morphism $\mathcal{O}'_Y \to \phi_* \mathcal{O}_X$ associated with Ψ is a monomorphism.

Θ is then characterised by the subsheaf (of abelian groups) $I = \mathrm{Ker}(\mathcal{O}_Y \to \mathcal{O}'_Y)$ of $\mathcal{O}_Y$, which is a sheaf of ideals in $\mathcal{O}_Y$: that is, equivalently

$$\forall y \in Y \ I_y \text{ is an ideal in } \mathcal{O}_{Y,y}$$

and $\forall$ open V in Y $\Gamma(V, I)$ is an ideal in $\Gamma(V, \mathcal{O}_Y)$.

Proof. Set $\mathcal{O}'_Y = \mathrm{SIm}(\mathcal{O}_Y \to \phi_* \mathcal{O}_X)$ (3.6.1); this is easily seen to be a sheaf of R-algebras with the desired properties, since $\mathcal{O}_Y \to \phi_* \mathcal{O}_X$ factors as $\mathcal{O}_Y \to \mathcal{O}'_Y \to \phi_* \mathcal{O}_X$. The exact sequence of sheaves of abelian groups

$$0 \to I \xrightarrow{f} \mathcal{O}_Y \xrightarrow{g} \mathcal{O}'_Y \to 0$$

shows that $I = \mathrm{Ker}(g)$ and $\mathcal{O}'_Y = \mathrm{SCok}(f)$ determine each other (3.6.6), and the stalk sequence over $y \in Y$ shows that I_y is an ideal of $\mathcal{O}_{Y,y}$. The equivalence of the second condition follows from the sheaf space interpretation (1.3). //

1.11 **Proposition.** If $\Theta : (X, \mathcal{O}_X) \to (Y, \mathcal{O}_Y)$ is a morphism of ringed spaces over R such that the underlying map $\theta : X \to Y$ is injective and the morphism $\mathcal{O}_Y \to \theta_* \mathcal{O}_X$ is a sheaf epimorphism, then Θ is a monomorphism in the category of ringed spaces over R.

Proof. We must show that given a diagram

$$(Z, \mathcal{O}_Z) \underset{\Psi}{\overset{\Phi}{\rightrightarrows}} (X, \mathcal{O}_X) \xrightarrow{\Theta} (Y, \mathcal{O}_Y)$$

such that $\Theta \circ \Phi = \Theta \circ \Psi$, we can deduce $\Phi = \Psi$. The diagram of underlying continuous maps is $Z \underset{\psi}{\overset{\phi}{\rightrightarrows}} X \xrightarrow{\theta} Y$ and since θ is injective, $\phi = \psi$. The two composites

$$\mathcal{O}_Y \to (\theta_* \mathcal{O}_X) \rightrightarrows (\theta \circ \phi)_* \mathcal{O}_Z$$

are equal by hypothesis. Applying 3. 7. 13 we get a diagram of sheaves on Z

$$(\theta \circ \phi)^*\mathcal{O}_Y = \phi^*\theta^*\mathcal{O}_Y \to \phi^*\mathcal{O}_X \rightrightarrows \mathcal{O}_Z$$

in which the composites are equal too. The stalk sequence over $z \in Z$ is

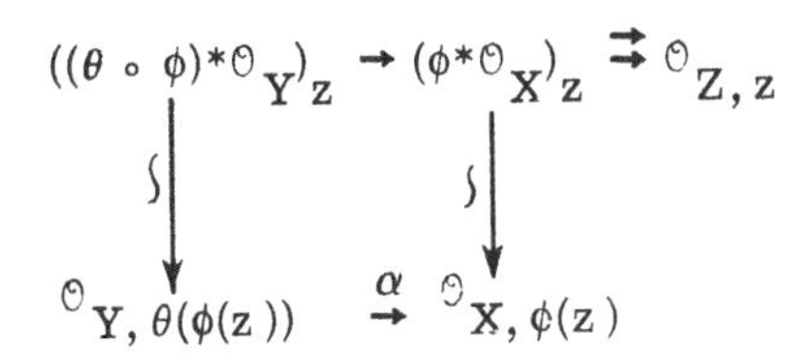

by 3. 7. 12, and by hypothesis α is surjective. Hence the two maps $\phi^*\mathcal{O}_X \rightrightarrows \mathcal{O}_Z$ are equal, and by 3. 7. 13 so are the maps $\mathcal{O}_X \rightrightarrows \phi_*\mathcal{O}_Z$ defining Φ and Ψ. Hence $\Phi = \Psi$. //

1. 12 **Scholium.** Any morphism $\Phi : (X, \mathcal{O}_X) \to (Y, \mathcal{O}_Y)$ of ringed spaces over R can be factored as

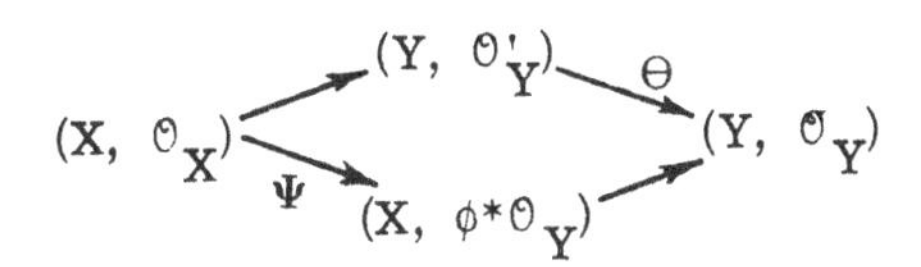

such that Θ is a monomorphism with underlying continuous map id_Y and Ψ has underlying continuous map id_X and is final among those ringed space structures on X through which Φ factors.

1. 13 **Hard exercise.** Prove or refute: the ringed space $(Y, \mathcal{O}'_Y)$ of 1. 10 and 1. 12 is the image of Φ in the category of ringed spaces over R.

For definiteness we record here the definition of image in an arbitrary category (cf. 3. 6. 2). The <u>image</u> of $f : A \to B$ is the smallest subobject of B through which f factors: that is, a factorisation

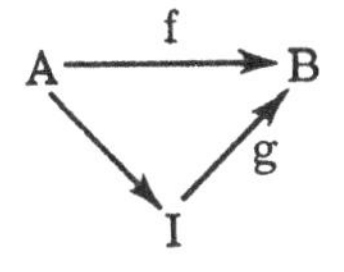

with g monomorphic, having the property that if

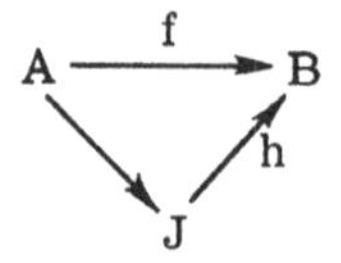

is another factorisation with h monomorphic, then g factors (necessarily uniquely) through h as

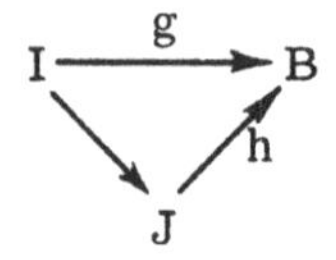

1.14 **Example.** Let $X = \{x\}$ be a one point space and $Y = \mathbf{R}^2$ be the real plane. Give X and Y the structure sheaves $\mathcal{O}_X$, $\mathcal{O}_Y$ of continuous **R**-valued functions (1.5B), and let $\phi : X \to Y$ map x to the point $y_0 \in Y$. The associated morphism of ringed spaces over **R** (1.8) has stalk map

$$\mathcal{O}_{Y,y_0} \to (\phi_* \mathcal{O}_X)_{y_0} \cong \mathcal{O}_{X,x} \cong \mathbf{R}$$

assigning to the germ of a function f its value $f(y_0)$ at y_0.

The sheaf $\mathcal{O}'_Y$ of 1.10 has stalks

$$\mathcal{O}'_{Y,y} = \begin{cases} \mathbf{R} & \text{if } y = y_0 \\ \{0\} & \text{if } y \neq y_0 \end{cases}$$

while the sheaf of ideals I of 1.10 has stalks

$$I_y = \mathcal{O}_{Y,y} \quad \text{if } y \neq y_0$$

and

$$I_{y_0} = \{g \in \mathcal{O}_{Y,y_0} ; \ g \text{ is the germ of a function } f \text{ such that } f(y_0) = 0\}$$

The sheaf $\phi^*\mathcal{O}_Y$ on X of 1.9 has stalk

$$(\phi^*\mathcal{O}_Y)_x = \mathcal{O}_{Y,y_0}$$

by 3.7.12 and the map $\mathcal{O}_{Y,y_0} \to \mathcal{O}_{X,x} \cong \mathbf{R}$ is again given by taking the value at y_0.

1.15 **Exercise.** Analyse an example like 1.14, but where ϕ is not injective.

4.2 The prime spectrum of a ring

2.1 Let R be a commutative ring with a one. We can associate with R a ringed space $(\operatorname{Spec} R, \mathcal{O})$ over R (or over $\mathbf{Z}$) as follows.

An ideal $\mathfrak{p}$ of R is called <u>prime</u> iff the ring $R/\mathfrak{p}$ is integral (= has no zero divisors and is not the zero ring). Let $\operatorname{Spec} R$ be the set of prime ideals of R, and for any ideal $\mathfrak{a}$ of R let

$$V(\mathfrak{a}) = \{ \mathfrak{p} \in \operatorname{Spec} R;\ \mathfrak{p} \supseteq \mathfrak{a} \}.$$

Then the $V(\mathfrak{a})$, as $\mathfrak{a}$ runs through all ideals of R, form the closed sets of a topology: this follows from the easy observations:

2.2 **Lemma.** (i) $V(\{0\}) = \operatorname{Spec} R$; $V(R) = \emptyset$.

(ii) $V(\Sigma_i \mathfrak{a}_i) = \cap_i V(\mathfrak{a}_i)$.

(iii) $V(\mathfrak{a}\mathfrak{b}) = V(\mathfrak{a}) \cup V(\mathfrak{b})$.

Proof. The only difficulty is $V(\mathfrak{a}\mathfrak{b}) \subseteq V(\mathfrak{a}) \cup V(\mathfrak{b})$; but if $\mathfrak{p} \in (\operatorname{Spec} R)\setminus(V(\mathfrak{a}) \cup V(\mathfrak{b}))$ then $\exists f \in \mathfrak{a} \setminus \mathfrak{p}$ and $\exists g \in \mathfrak{b} \setminus \mathfrak{p}$ so that $fg \notin \mathfrak{p}$ but $fg \in \mathfrak{a}\mathfrak{b}$; therefore $\mathfrak{p} \not\supseteq \mathfrak{a}\mathfrak{b}$. //

2.3 For $f \in R$, let

$$D(f) = \{ \mathfrak{p} \in \operatorname{Spec} R;\ f \notin \mathfrak{p} \} = \operatorname{Spec} R \setminus V(Rf).$$

Then the $D(f)$ for $f \in R$ form a basis for the topology on $X = \operatorname{Spec} R$; for if $\mathfrak{p} \in X \setminus V(\mathfrak{a})$ then picking $f \in \mathfrak{a} \setminus \mathfrak{p}$, we have $\mathfrak{p} \in D(f) \subseteq X \setminus V(\mathfrak{a})$.

Also $D(f) \cap D(g) = D(fg)$, so that the basis $\{D(f);\ f \in R\}$ is closed under finite intersections.

2.4 **Examples.**

A. If k is any field, $\operatorname{Spec} k$ has just one point, $\{0\}$.

B. Let k be an algebraically closed field (e.g. $\mathbf{C}$), and let $R = k[x]$ be the polynomial ring in one variable. The only prime ideals of R are $\{0\}$ and the $R(x - \alpha)$ for $\alpha \in k$. Hence $\operatorname{Spec}(k[x]) = \{\{0\}\} \cup M$

where $M \xrightarrow{\text{bij}} k$; the topology induced on M has the finite sets as its closed sets, and in Spec R the closure of the point $\{0\}$ is the whole space.

C. If k is an algebraically closed field, then

$$\text{Spec } k[x_1, \ldots, x_n] \supseteq M$$

where M is the set of closed points (= maximal ideals), and by the Weak Nullstellensatz $M \xrightarrow{\text{bij}} k^n$, since any maximal ideal is of the form $(x_1 - \alpha_1, x_2 - \alpha_2, \ldots, x_n - \alpha_n)$ for some $\alpha_i \in k$ $(1 \le i \le n)$.

The topology induced on k^n is the Zariski topology, having as its closed sets all the sets of common zeros of a collection of polynomials: that is, all sets of the form

$$\{(\alpha_1, \ldots, \alpha_n) \in k^n; \quad \forall 1 \le i \le m \quad f_i(\alpha_1, \alpha_2, \ldots, \alpha_n) = 0\}$$

where $f_1, \ldots, f_m \in k[x_1, \ldots, x_n]$.

D. Spec $\mathbf{Z}$ has one closed point for each prime number $p \in \mathbf{N}$, and one other point.

2.5 To each basic open set of Spec R we can assign a ring:

$$D(f) \mapsto R_f$$

(where R_f is the ring of fractions of R with respect to the multiplicative system $\{1, f, f^2, \ldots\}$), and to any containment $D(f) \supseteq D(g)$ a restriction ring morphism $R_f \to R_g$; namely

$$D(f) \supseteq D(g) \iff \exists n \in \mathbf{N},\ a \in R \text{ such that } g^n = af;$$

then define

$$R_f \to R_g : \frac{r}{f^m} \mapsto \frac{a^m r}{g^{nm}}$$

(mnemonic for this:

$$\text{"}\ \frac{r}{f^m} = \frac{a^m r}{(af)^m} = \frac{a^m r}{g^{mn}}\ \text{"}\ ).$$

In particular, if $D(f) = D(g)$, then $R_f \xrightarrow{\sim} R_g$, so that the assignment $D(f) \mapsto R_f$ is 'well-defined'. (See also [EGAI] 1.3.1.)

Since $D(f) \cap D(g) = D(fg)$, if $D(f) = \cup_{\lambda \in \Lambda} D(f_\lambda)$ is a cover of a basic open set by basic open sets, the sequence of 2.1.6 becomes

$$R_f \to \Pi_{\lambda \in \Lambda} R_{f_\lambda} \rightrightarrows \Pi_{(\lambda, \mu) \in \Lambda \times \Lambda} R_{f_\lambda f_\mu}$$

and we can check by commutative algebra that this sequence is an equaliser. (For details, see Macdonald, Algebraic geometry, Prop. 5.1 or Mumford, Introduction to algebraic geometry, Chapter 2, §1.)

Hence we have the data for a 'sheaf defined only on the open sets of a basis', and we need the following sheaf-theoretic Lemma.

2.6 **Lemma.** Suppose that X is a topological space and $\mathfrak{U}$ is a basis for the topology which is closed under finite intersections. Let F be the data of a presheaf (sets of sections and restriction maps) given only for open sets of the basis, which satisfies the condition that whenever $U = \cup_{\lambda \in \Lambda} U_\lambda$ with $U \in \mathfrak{U}$ and $\forall \lambda \in \Lambda$ $U_\lambda \in \mathfrak{U}$, the sequence

$$F(U) \to \Pi_{\lambda \in \Lambda} F(U_\lambda) \rightrightarrows \Pi_{(\lambda, \mu) \in \Lambda \times \Lambda} F(U_\lambda \cap U_\mu)$$

is an equaliser diagram (maps as in 2.1.6).

Then there is a sheaf G on X, unique up to isomorphism, such that

$$\forall U \in \mathfrak{U} \qquad \Gamma(U, G) = F(U)$$

and $\forall U, V \in \mathfrak{U}$, if $U \supseteq V$ the restriction maps $F(U) \to F(V)$ and $\Gamma(U, G) \to \Gamma(V, G)$ agree.

Sketch of one possible proof:

Step 1: If such a G exists, it must have stalks

$$G_x = \varinjlim_{\mathfrak{U} \ni U \ni x} F(U)$$

so that the underlying set of the sheaf space LG is determined.

Step 2: The topology on LG is also determined, since the $\hat{s}[U]$ with

$U \in \mathfrak{U}$ and $s \in F(U)$ must form a basis for it (cf. 2.3.8 and 2.3.5). Hence the sheaf G is unique if it exists.

Step 3: Construct G by forming the stalks $G_x = \varinjlim_{\mathfrak{U} \ni U \ni x} F(U)$ and thus a sheaf space $E = \coprod_{x \in X} G_x$, and check as in 2.4.3 that $F(U) \xrightarrow{\sim} \Gamma(U, E)$ for $U \in \mathfrak{U}$. //

An alternative construction is to set, for V open in X,

$$\Gamma(V, G) = \varprojlim_{V \supseteq U \in \mathfrak{U}} F(U).$$

2.7 Applying Lemma 2.6 to the situation of 2.5 we see that $X = \operatorname{Spec} R$ has on it a sheaf of rings, denoted by $\mathcal{O}$ or $\mathcal{O}_X$, with the property that for $f \in R$

$$\Gamma(D(f), \mathcal{O}) = R_f.$$

In particular, putting $f = 1$ we have

$$\Gamma(X, \mathcal{O}) = R$$

so that we can recover the ring R from the ringed space $(\operatorname{Spec} R, \mathcal{O})$ (at least up to isomorphism).

At the point $\mathfrak{p} \in \operatorname{Spec} R$, $\mathcal{O}$ has stalk

$$\mathcal{O}_{\mathfrak{p}} = \varinjlim_{D(f) \ni \mathfrak{p}} R_f = \varinjlim_{f \notin \mathfrak{p}} R_f$$

$$= R_{\mathfrak{p}} \qquad \text{(by commutative algebra),}$$

where $R_{\mathfrak{p}}$ is the localisation $(R \setminus \mathfrak{p})^{-1}R$ of R at the prime ideal $\mathfrak{p}$. Rings of the form $R_{\mathfrak{p}}$ are all <u>local</u> rings; that is, they each have a unique maximal ideal. In fact, as the stalk of the structure sheaf $\mathcal{O}$ at the point $\mathfrak{p}$, $R_{\mathfrak{p}}$ describes the nature of the ringed space $\operatorname{Spec} R$ 'near' the point $\mathfrak{p}$, and this is the origin of the term 'local ring'.

2.8 **Definition.** A ringed space (over $\mathbf{Z}$) isomorphic to $(\operatorname{Spec} R, \mathcal{O})$ for some ring R is called an <u>affine scheme.</u> As the construction of 2.5 and 2.7 shows, $(\operatorname{Spec} R, \mathcal{O})$ has a natural structure of ringed space over R; indeed for any S such that R is an S-algebra (that is, we are given a ring morphism $S \to R$), Spec R is a ringed space over S.

2.9 Examples.

A. Although for all fields k the topological spaces Spec k are homeomorphic, they are distinguished by their structure sheaves, since $k \cong \Gamma(\operatorname{Spec} k, \mathcal{O}) \cong \mathcal{O}_P$ where P is the point of Spec k.

B. For any ring R, the affine scheme $\operatorname{Spec} R[x_1, x_2, \ldots, x_n]$ of the polynomial ring in n variables is called affine n-space over R, and is denoted by $\mathbf{A}^n_R$.

For $R = k$, an algebraically closed field, as in 2.4B and C, and $S = k[x_1, \ldots, x_n]$, all the S_f (for $f \in S$) can be considered as subrings of

$$k(x_1, \ldots, x_n) = \text{the field of fractions of } S.$$

At a point $(x_1 - \alpha_1, \ldots, x_n - \alpha_n)$ of $M \subseteq \operatorname{Spec} S$, $f \in S$ has the value $f(\alpha_1, \ldots, \alpha_n) \in k$; this corresponds to taking the image of f in the residue field $R_\mathfrak{p}/\mathfrak{p}R_\mathfrak{p}$ of the local ring at $\mathfrak{p} = (x_1 - \alpha_1, \ldots, x_n - \alpha_n)$. Then

$$M \cap D(f) \cong \{(\alpha_1, \ldots, \alpha_n) \in k^n; \quad f(\alpha_1, \ldots, \alpha_n) \neq 0\}$$

and for $s \in k(x_1, \ldots, x_n)$ and U open in $\operatorname{Spec} S = \mathbf{A}^n_k$ we have

$$s \in \Gamma(U, \mathcal{O}) \iff \forall (\alpha_1, \ldots, \alpha_n) \in M \cap U \quad s \text{ can be written in the form } s = \frac{f}{g} \text{ with } f, g \in S \text{ and } g(\alpha_1, \ldots, \alpha_n) \neq 0.$$

Thus if we look just at $\mathcal{O}|M$, we see the sheaf of algebraic functions on k^n. Compare 2. Ex. 6.

2.10 Proposition. Spec is a contravariant functor: that is, a ring morphism $r : R \to S$ gives rise in a natural way to a continuous map

$$\phi : \operatorname{Spec} S \to \operatorname{Spec} R$$

and a ϕ-morphism of sheaves of rings $\mathcal{O}_{\operatorname{Spec} R} \to \mathcal{O}_{\operatorname{Spec} S}$. Furthermore for each $x \in \operatorname{Spec} S$ the stalk morphism

$$\mathcal{O}_{\operatorname{Spec} R, \phi(x)} \to \mathcal{O}_{\operatorname{Spec} S, x}$$

is a local morphism of local rings; that is, it takes the maximal ideal into the maximal ideal.

Proof. If $\mathfrak{p}$ is a prime ideal of S, let $\phi(\mathfrak{p}) = r^{-1}[\mathfrak{p}]$; then since the ring morphism

$$R/r^{-1}[\mathfrak{p}] \to S/\mathfrak{p}$$

is injective, $\phi(\mathfrak{p})$ is a prime ideal of R. Since for $f \in R$

$$\begin{aligned} \phi^{-1}[D(f)] &= \{ \mathfrak{p} \in \operatorname{Spec} S;\ \phi(\mathfrak{p}) \in D(f) \} \\ &= \{ \mathfrak{p} \in \operatorname{Spec} S;\ r^{-1}[\mathfrak{p}] \not\ni f \} = D(rf) \end{aligned}$$

the map ϕ is continuous. Under the map

$$R \xrightarrow{r} S \to S_{r(f)}$$

f becomes a unit, so we get induced maps

$$\Gamma(D(f), \mathcal{O}) = R_f \to S_{rf} = \Gamma(D(rf), \mathcal{O})$$

giving the ϕ-morphism required. Similarly, since for $\mathfrak{p} \in \operatorname{Spec} S$

$$R \xrightarrow{r} S \to S_{\mathfrak{p}}$$

sends every $f \notin \phi(\mathfrak{p})$ into a unit, the induced map

$$R_{\phi(\mathfrak{p})} \to S_{\mathfrak{p}}$$

of stalks sends each member of the maximal ideal $\phi(\mathfrak{p}).R_{\phi(\mathfrak{p})}$ into the maximal ideal $\mathfrak{p} S_{\mathfrak{p}}$. //

2.10 **Scholium.** A ring morphism $R \to S$ makes S into an R-algebra, and induces a morphism $\operatorname{Spec} S \to \operatorname{Spec} R$ of ringed spaces over R. We may choose to regard it as a morphism of ringed spaces over $\mathbf{Z}$, since each of R, S is a $\mathbf{Z}$-algebra in a unique way (cf. 1.1, 1.2 and 2.8).

4.3 Geometric spaces and manifolds

3.1 **Definition.** Let R be a ring. A ringed space $(X, \mathcal{O}_X)$ over R is called a geometric space (over R) iff all the stalks $\mathcal{O}_{X,x}$ (for $x \in X$) are local rings (that is, each has a unique maximal ideal, called m_x).

A morphism of geometric spaces over R $(X, \mathcal{O}_X) \to (Y, \mathcal{O}_Y)$ (or R-morphism) is a morphism of ringed spaces, with underlying continuous map $\phi : X \to Y$ and ϕ-morphism of sheaves of R-algebras $\psi : \mathcal{O}_Y \to \mathcal{O}_X$, with the additional property that for each $x \in X$ the stalk map

$$\psi_x : \mathcal{O}_{Y,\phi(x)} \to \mathcal{O}_{X,x}$$

is a local morphism of local rings; that is $\psi_x(m_{\phi(x)}) \subseteq m_x$.

Thus we have defined a (non-full) subcategory of the category of ringed spaces over R.

3.2 **Examples.**

A. By 2.7 and 2.8 any affine scheme Spec R is a geometric space (over R or **Z**), and any ring morphism $R \to S$ gives rise by 2.10 to a morphism of geometric spaces Spec S $\to$ Spec R (again over R or **Z**).

B. Let R be any ring, and F an R-algebra which is also a topological field (so that the operations of subtraction and division are continuous); for instance $R = F = \mathbf{R}$, or $R = \mathbf{Z}$, $F = \mathbf{C}$. For any topological space X, let C^F be the sheaf of continuous F-valued functions on X (cf. 1.5B). Then (X, C^F) is a geometric space over R (see 3.3).

C. The example (X, C^r) of 1.5C (with X a banach space over **R**) is a geometric space over **R** (see 3.3).

D. The example (X, C^ω) of 1.5D is a geometric space over K (see 3.3).

3.3 In each of the cases 3.2B, 3.2C, 3.2D the stalks $\mathcal{O}_{X,x}$ are local rings with maximal ideal

$$m_x = \{g \in \mathcal{O}_{X,x};\ g \text{ is the germ of a function } f \text{ such that } f(x)=0\}.$$

For this is clearly an ideal, and $\mathcal{O}_{X,x}/\mathfrak{m}_x \cong F$ or $\mathbf{R}$ or K (according to the case) by $g \mapsto$ value of g at x, so that $\mathfrak{m}_x$ is maximal (since F, $\mathbf{R}$, K are each fields). Also $g \notin \mathfrak{m}_x$ implies that g is the germ of a function f such that $f(x) \neq 0$, and so in a small enough neighbourhood of x, $\frac{1}{f}$ exists; hence g is a unit in $\mathcal{O}_{X,x}$, so that $\mathfrak{m}_x$ is the unique maximal ideal of $\mathcal{O}_{X,x}$.

Note that in each case the R-algebra structure on $\mathcal{O}_X$ is given by the maps $R \to \Gamma(U, \mathcal{O}_X)$ (for U open in X) which send $r \in R$ to the constant function with value r.

3.4 When we look at geometric space morphisms between geometric spaces of the types of 3.2B, C, D, a remarkable thing happens. Let us examine for example the typical case of an $\mathbf{R}$-morphism $(X, \mathcal{O}_X) \to (Y, \mathcal{O}_Y)$ where

$X = \mathbf{R}^n$, $\mathcal{O}_X = C^r$ is the sheaf of differentiable functions on X;
$Y = \mathbf{R}^m$, $\mathcal{O}_Y = C^r$ is that on Y.

The underlying continuous map $\phi : \mathbf{R}^n \to \mathbf{R}^m$ is given by say

$$\phi(x) = (\rho_1(x), \ldots, \rho_m(x)) \in \mathbf{R}^m$$

where if $\pi_i : Y \to \mathbf{R}$ is projection onto the ith coordinate, we have

$$\rho_i = \pi_i \circ \phi.$$

Now consider the ϕ-morphism $\psi : \mathcal{O}_Y \to \mathcal{O}_X$. This has the property that each stalk map

$$\mathcal{O}_{Y,\phi(x)} \to \mathcal{O}_{X,x} \quad \text{for } x \in X$$

is local; hence by the above interpretation of the maximal ideals, if f is a differentiable function on some open set V in Y, we have

$$\forall x \in \phi^{-1}(V) \quad f(\phi(x)) = 0 \Rightarrow \psi(f)(x) = 0$$

$$\text{(i.e. } f_{\phi(x)} \in \mathfrak{m}_{\phi(x)}) \qquad \text{(i.e. } \psi_x(f_{\phi(x)}) \in \mathfrak{m}_x).$$

Applying this to the differentiable function $(f - c)$ with $c \in \mathbf{R}$ we get

$$\forall x \in \phi^{-1}(V) \quad f(\phi(x)) = c \Rightarrow \psi(f)(x) = c \quad (*34);$$

in other words $\psi(f)(x) = f(\phi(x))$, so that

$$\psi(f) = f \circ \phi.$$

Thus the sheaf morphism ψ is just 'compose with ϕ':

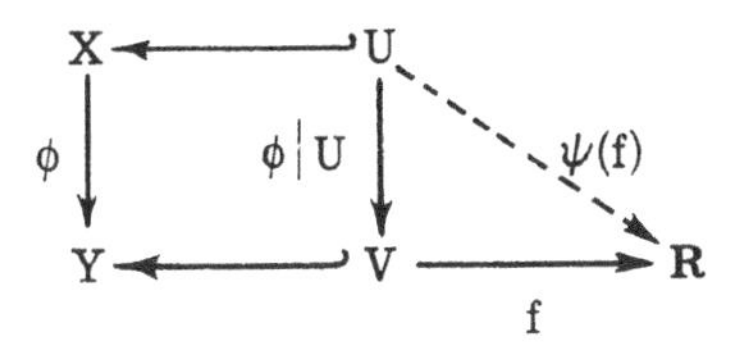

as in 1.8. In particular, $\psi(\pi_i) = \pi_i \circ \phi = \rho_i$; but each π_i is differentiable and so each $\psi(\pi_i) = \rho_i$ is differentiable, so that ϕ is actually an (r times continuously) differentiable map: $\mathbf{R}^n \to \mathbf{R}^m$.

Thus the effect of insisting that the a priori merely continuous ϕ has an associated ϕ-morphism $C^r_Y \to C^r_X$ is to ensure that ϕ is actually differentiable. Conversely, a differentiable map $\phi : X \to Y$ clearly defines a ϕ-morphism $C^r_Y \to C^r_X$ by the above procedure of composing with ϕ, and hence a geometric space morphism: $X \to Y$.

A similar argument shows that any K-morphism of geometric spaces $(X, C^\omega) \to (Y, C^\omega)$ (spaces as in 3.2D) has an underlying map which is analytic.

In each case the step (*34) is valid, since the morphism is over $R = \mathbf{R}$ or K and so takes the constant function with value $r \in R$ to the constant function with value r.

3.5 **Exercise.** Show that any geometric space morphism between affine schemes Spec S → Spec R arises from a ring morphism R → S as in 2.10. Deduce that the category of affine schemes and geometric space morphisms is contravariant-equivalent to the category of rings.

3.6 **Definition.** Fix a ring R. Suppose we have a geometric space $(M, \mathcal{O}_M)$ over R (to be regarded as a 'model'). We say that a geometric space $(X, \mathcal{O}_X)$ over R is <u>locally isomorphic</u> to $(M, \mathcal{O}_M)$ iff $\forall x \in X$ there is an open neighbourhood U of x in X and an open set V in M such that there is an isomorphism

$$(U, \mathcal{O}_X|U) \cong (V, \mathcal{O}_M|V)$$

as geometric spaces over R.

Given a class $\mathfrak{M}$ of model spaces $(M, \mathcal{O}_M)$, we say that $(X, \mathcal{O}_X)$ is a manifold of type $\mathfrak{M}$ (over R) iff X can be covered by open sets U such that each $\mathcal{O}_X|U$ is locally isomorphic to some $(M, \mathcal{O}_M) \in \mathfrak{M}$.

A morphism between manifolds of type $\mathfrak{M}$ is just a morphism of geometric spaces over R.

Some authors use the word variety in place of manifold.

3.7 Examples.

A. Let $R = \mathbf{R}$.

Putting $\mathfrak{M} = \{(\mathbf{R}^n, C^{\mathbf{R}})\}$ we obtain topological manifolds of dimension n.

Putting $\mathfrak{M} = \{(\mathbf{R}^n, C^r)\}$ we obtain differentiable manifolds of class r and dimension n.

Putting $\mathfrak{M} = \{(\mathbf{R}^n, C^\omega); n \in \mathbf{N}\}$ we obtain real analytic manifolds.

B. Let $R = \mathbf{C}$.

Putting $\mathfrak{M} = \{(\mathbf{C}, C^\omega)\}$ we obtain Riemann surfaces.

Putting $\mathfrak{M} = \{(\mathbf{C}^n, C^\omega); n \in \mathbf{N}\}$ we obtain complex analytic manifolds.

C. Let $R = \mathbf{Q}_p$.

Putting $\mathfrak{M} = \{(\mathbf{Q}_p^n, C^\omega); n \in \mathbf{N}\}$ we obtain p-adic analytic manifolds.

Putting $\mathfrak{M} = \{(E, C^\omega); E$ is a hilbert space over $\mathbf{Q}_p\}$ we obtain p-adic analytic manifolds of hilbert type.

D. Let $R = \mathbf{Z}$.

Putting $\mathfrak{M} =$ class of all affine schemes, we obtain schemes.

3.8 Remark. After 3.4 and 3.5 each of the above definitions is in accordance with the more usual definitions in terms of atlases of charts with transition maps of the appropriate kind (see the references for 3.9 below), with two possible exceptions. Some authors may require that X have a countable base of open sets. Other authors may insist that $(X, \mathcal{O}_X)$ satisfy a separation (hausdorff) condition of some kind to avoid

examples like

(= $\mathbf{R} \sqcup \mathbf{R}/\sim$ where '$x \sim y$ iff $x = y$ and $x \neq 0$'). This condition is just that the underlying space X be a hausdorff space in 3.7A, B, C but is a little more complicated in the case 3.7D of schemes (it gives separated schemes).

3.9 **Exercise.** Compare 3.6 and 3.7 with any other definition of manifold you may have met; for example as in Bourbaki, Variétés différentielles et analytiques; [H] Chapter I, 2.5; [L] Chapter 2.

In particular, verify (using 3.4) that the geometric space morphisms between topological (respectively differentiable, analytic) manifolds are just the continuous (respectively differentiable, analytic) maps that are usually considered between manifolds. What about piece-wise-linear (PL) manifolds?

3.10 **Aside.** In proving in 3.4 that the geometric space morphisms between geometric spaces of types 3.2C and 3.2D were precisely the maps we desired between the underlying spaces, it was vital that the sheaf morphism ψ preserved constant functions (*34). We used the supposition that ψ was a morphism of sheaves of R-algebras to justify this. However, it is an elementary fact of analysis that for $R = \mathbf{R}$ any morphism of sheaves of rings preserves all the ($\mathbf{R}$-valued) constant functions. This follows from the Lemma below, which we prove for the convenience of the reader. For $\mathbf{R}$-valued functions f, g on a set U we write

$$f > g \iff \forall x \in U \quad f(x) > g(x)$$

and we denote the constant function with value $c \in \mathbf{R}$ again by c.

Lemma. Let U, V be sets and let R (resp. S) be a subring of the ring of $\mathbf{R}$-valued functions on U (resp. V) such that:

(i) $\forall c \in \mathbf{R} \quad c \in R$ (resp. S)

(ii) $\forall f \in R$ (resp. S) $f > 0 \Rightarrow \exists g \in R$ (resp. S) such that $f = g^2$.

Then if $\psi : \mathbf{R} \to S$ is any ring morphism, we have

$$\forall c \in \mathbf{R} \quad \psi(c) = c.$$

Proof. Since $\psi(1) = 1$ we have for $n \in \mathbf{N}$

$$\psi(n) = \psi(\textstyle\sum_{i=1}^{n} 1) = \sum_{i=1}^{n} \psi(1) = n,$$

and since $\psi(-n) + \psi(n) = 0$, we have $\psi(-n) = -n$. For $m, n \in \mathbf{Z}$

$$n\psi(\tfrac{m}{n}) = \psi(m) = m$$

so $\forall q \in \mathbf{Q} \quad \psi(q) = q$.

We have for $f \in \mathbf{R}$

$$\begin{aligned} f > 0 &\iff \exists g \in \mathbf{R} \text{ such that } f = g^2 \\ &\Rightarrow \exists g \in \mathbf{R} \text{ such that } \psi(f) = (\psi(g))^2 \\ &\Rightarrow \psi(f) \geq 0. \end{aligned}$$

Now suppose we are given $c \in \mathbf{R}$. For any $\varepsilon > 0 \; \exists \; q, q' \in \mathbf{Q}$ with

$$q > c > q' \quad \text{and} \quad q - q' < \varepsilon$$

and then $\psi(q - c) \geq 0$, $\psi(c - q') \geq 0$ so $q \geq \psi(c) \geq q'$ and $q - q' < \varepsilon$. Letting $\varepsilon \to 0$ we see that $\psi(c) = c$. //

Corollary. $\mathbf{R}$ has no non-trivial ring automorphism.

Proof. Let $U = V$ have one point. //

We note that condition (ii) is satisfied for the subrings of differentiable (or analytic) functions on $\mathbf{R}^n$ since we can compose with the square root function which is analytic away from zero.

Hence if we were interested only in geometric spaces and manifolds over $\mathbf{R}$, we need only have insisted that the structure sheaves be sheaves of rings (that is $\mathbf{Z}$-algebras) in order to get the 'correct' notion of morphism.

However this is no longer true if we wish to consider other types of manifold. For instance if σ is any automorphism of $\mathbf{C}$ (for instance complex conjugation), then for an analytic $f : \mathbf{C} \to \mathbf{C}$, $\sigma \circ f \circ \sigma^{-1}$ is again analytic, and $f \mapsto \sigma \circ f \circ \sigma^{-1}$ is an automorphism of the ring of analytic $\mathbf{C}$-valued functions on $\mathbf{C}$ which does not preserve the constant functions.

3.11 The Spec construction has a universal property among all geometric spaces.

Theorem. *Let* $(X, \mathcal{O}_X)$ *be a geometric space over* $\mathbf{Z}$, *and let* R *be a ring. For each ring morphism* $t : R \to \Gamma(X, \mathcal{O}_X)$ *there is a unique morphism of geometric spaces*

$$G : (X, \mathcal{O}_X) \to (\operatorname{Spec} R, \mathcal{O})$$

such that the induced map

$$R \xrightarrow{\sim} \Gamma(\operatorname{Spec} R, \mathcal{O}) \to \Gamma(X, \mathcal{O}_X)$$

is t. *In other words the map*

$$\operatorname{Hom}_{Gsp}((X, \mathcal{O}_X), (\operatorname{Spec} R, \mathcal{O})) \to \operatorname{Hom}_{Ring}(R, \Gamma(X, \mathcal{O}_X)) : \Phi \mapsto {}'\Gamma(X, \Phi)'$$

is bijective (where Gsp *is the category of geometric spaces over* $\mathbf{Z}$).

Proof. Given t, we first construct the underlying continuous map g of G. For $x \in X$, let t_x be defined to make the diagram

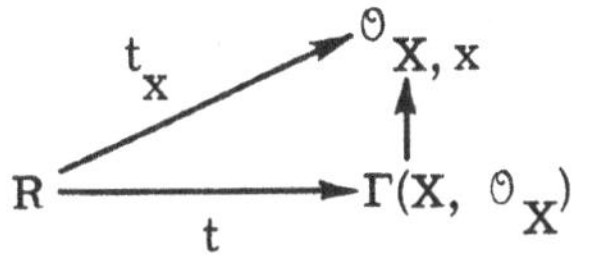

$$\begin{array}{ccc} & & \mathcal{O}_{X,x} \\ & \overset{t_x}{\nearrow} & \uparrow \\ R & \xrightarrow{\quad t \quad} & \Gamma(X, \mathcal{O}_X) \end{array}$$

commute. Let $g(x) = t_x^{-1}(m_x)$ where m_x is the maximal ideal of $\mathcal{O}_{X,x}$; then $g(x) \in \operatorname{Spec}(R)$. For $f \in R$ we have

$$\begin{aligned} g^{-1}(D(f)) &= \{x \in X;\ g(x) \not\ni f\} \\ &= \{x \in X;\ t_x(f) \notin m_x\} \end{aligned}$$

and this is open in X, for $f' = t_x(f) \notin m_x$ holds iff f' is a unit in $\mathcal{O}_{X,x}$, so $\exists g' \in \mathcal{O}_{X,x}$ such that $f'.g' = 1$, and this equality still holds in some neighbourhood of x.

This calculation also shows that the map

$$R \to \Gamma(X, \mathcal{O}_X) \to \Gamma(g^{-1}(D(f)), \mathcal{O}_X)$$

inverts f, so it factorises as

$$\Gamma(D(f), \mathcal{O}) = R_f \to \Gamma(g^{-1}(D(f)), \mathcal{O}_X)$$

and this defines the morphism G, which clearly has t as its associated global section map.

The uniqueness follows from the easily verified fact that this construction is inverse to the process of taking global sections (so that the map given in the Theorem is bijective). //

3.12 **Remark.** 3.11 shows that Spec and $\Gamma(-, \mathcal{O}_-)$ are adjoint functors. It follows that to give $(X, \mathcal{O}_X)$ the structure of a ringed space over R (that is to make $\Gamma(X, \mathcal{O}_X)$, and hence all the $\Gamma(U, \mathcal{O}_X)$ for U open in X, into R-algebras) is the same as to give a morphism of geometric spaces (over $\mathbf{Z}$) $X \to \operatorname{Spec} R$.

4.4 Modules over ringed spaces

4.1 Recall that if R is a ring with a one, a (left) R-module is an additive abelian group M equipped with a map

$$R \times M \to M : (r, m) \mapsto rm$$

(that is, an operation of R on M) satisfying the conditions:

(i) $\forall r \in R\ \forall m, n \in M \qquad r(m + n) = rm + rn$

(ii) $\forall r, s \in R\ \forall m \in M \qquad (r + s)m = rm + sm$ and $(rs)m = r(sm)$

(iii) $\forall m \in M \qquad 1m = m.$

A morphism $f : M \to N$ of R-modules is an abelian group morphism which satisfies the additional condition:

$$\forall m \in M\ \forall r \in R \qquad f(rm) = rf(m)$$

(so that f respects (or preserves) the operation of R).

The category R-mod of all R-modules is abelian (cf. 3.5.4) (even if R is not commutative); indeed, the kernels, cokernels and biproducts are constructed as if in the category Abgp, and are then seen to carry a natural R-module structure.

For example, for R a field we get the category of all R-vector spaces; while for $R = \mathbf{Z}$ we get the category of all abelian groups (for to prescribe $a - b$ is to prescribe na for all $n \in \mathbf{Z}$).

4.2 More generally, if $\rho : R \to S$ is a (one-preserving) ring morphism, and M is an R-module and N is an S-module, then a map $f : M \to N$ is called a morphism over ρ (or dihomomorphism) iff f is an abelian group morphism and

$$\forall r \in R \quad \forall m \in M \qquad f(rm) = \rho(r)f(m).$$

4.3 **Example.** If R' is an R-algebra, with structure map $\alpha : R \to R'$, then R' is an R-module with the operation of R defined by

$$rr' = \alpha(r).r' \quad \text{(product in } R'\text{) for } r \in R,\ r' \in R'.$$

A commutative square of ring morphisms:

$$\begin{array}{ccc} R' & \xrightarrow{\rho'} & S' \\ \alpha \uparrow & & \uparrow \beta \\ R & \xrightarrow{\rho} & S \end{array}$$

may be regarded as giving a module morphism $\rho' : R' \to S'$ over ρ.

4.4 **Definition.** Suppose that $(X, \mathcal{O})$ is a ringed space over a ring R. An $\mathcal{O}$-Module (note the use of the capital letter), or sheaf of $\mathcal{O}$-modules, M is a sheaf of abelian groups on X with the additional structure that

(i) for each open U in X, $\Gamma(U, M)$ is a $\Gamma(U, \mathcal{O})$-module; and

(ii) whenever $V \subseteq U$ are open in X the restriction map

$$\Gamma(U, M) \to \Gamma(V, M)$$

is a module morphism over the ring morphism (restriction) $\Gamma(U, \mathcal{O}) \to \Gamma(V, \mathcal{O})$ (as in 4.2).

There is an obvious definition of morphism for $\mathcal{O}$-Modules: $f : M \to N$ must be a morphism of abelian sheaves, and for each open U in X

$$\Gamma(U, f) : \Gamma(U, M) \to \Gamma(U, N)$$

must be a $\Gamma(U, \mathcal{O})$-module morphism.

Hence we have defined a category $\mathcal{O}$-Mod of all $\mathcal{O}$-Modules on X.

4.5 Examples.

A. For any topological space X, let $\mathcal{O}$ be the constant sheaf $\mathbf{Z}$ on X (as in 1.5A). Then any abelian sheaf on X is an $\mathcal{O}$-Module, and the category $\mathcal{O}$-Mod is just the category Shv/X we dealt with in Chapter 3.

B. Let R be a ring, X = Spec R and $\mathcal{O}$ be the structure sheaf on X as in 2.7. Starting with an R-module M, we can construct an $\mathcal{O}$-Module $\tilde{M}$ on Spec R as follows.

To each basic open set D(f) of Spec R we associate the R_f-module $M_f = S^{-1}M$ where $S = \{f^n;\ n \in \mathbf{N}\}$. The restriction maps follow from Commutative Algebra, and the sequence

$$M_f \to \Pi_{\lambda \in \Lambda} M_{f_\lambda} \rightrightarrows \Pi_{(\lambda, \mu) \in \Lambda \times \Lambda} M_{f_\lambda f_\mu}$$

corresponding to an open cover $D(f) = \cup_{\lambda \in \Lambda} D(f_\lambda)$ is an equaliser.

Hence by Lemma 2.6, M defines a sheaf $\tilde{M}$ on Spec R which is easily seen to be an $\mathcal{O}$-Module.

We remark in passing that $\tilde{R} \cong \mathcal{O}$, and that it can be shown that the category R-mod is equivalent to a full subcategory of $\mathcal{O}$-Mod by the assignment $M \mapsto \tilde{M}$. An $\mathcal{O}_X$-Module on a scheme $(X, \mathcal{O}_X)$ which is given locally (on an open cover by affine schemes) by Modules of the form $\tilde{M}$ is called <u>quasi-coherent.</u>

C. Suppose that $(X, \mathcal{O}_X)$ is a differentiable or analytic manifold in the interpretation of 3.7. Then for $p \in \mathbf{N}$ the assignment

$$U \mapsto \text{set of p-forms on } U \quad (\text{for } U \text{ open in } X)$$

defines an $\mathcal{O}_X$-Module.

Similarly the sheaves of sections (cf. 2.2.C) of vector bundles over a topological, differentiable or analytic manifold $(X, \mathcal{O}_X)$ can be regarded as $\mathcal{O}_X$-Modules.

(References: [H] Ch. I, §§3.2, 3.6; [L] Ch. III, V; Atiyah, K-Theory, Ch. I.)

4.6 It is easy to see from the construction of direct limits that if M is an $\mathcal{O}$-Module on X and $x \in X$, then the stalk M_x has a natural $\mathcal{O}_x$-module structure.

Applying the definitions of kernel (3.3.1), cokernel (3.4.4) and biproduct (3.5.1) to $\mathcal{O}$-Modules and $\mathcal{O}$-Module morphisms, we get $\mathcal{O}$-Modules and morphisms again. All the results of §§3.2-3.6 hold for the category $\mathcal{O}$-Mod in place of Shv/X. In particular, $\mathcal{O}$-Mod is an abelian category, and a sequence $K \to L \to M$ in $\mathcal{O}$-Mod is exact iff it is exact as a sequence of sheaves of abelian groups (which holds iff each stalk sequence is exact, by 3.6.5).

4.7 **Moral** (after 4.5A). We should have been dealing with $\mathcal{O}$-Modules all along.

4.8 **Constructions.** Recall that if R is a ring and $(M_\lambda)_{\lambda\in\Lambda}$ a family of R-modules, then the direct product $\Pi_{\lambda\in\Lambda}M_\lambda$ is the R-module defined by giving the product set pointwise operations. The direct sum $\oplus_{\lambda\in\Lambda}M_\lambda$ is the sub-R-module of $\Pi_{\lambda\in\Lambda}M_\lambda$ generated by the images of all the injection maps

$$i_\mu : M_\mu \to \Pi_{\lambda\in\Lambda}M_\lambda : m \mapsto (m_\lambda)_{\lambda\in\Lambda}$$

where $m_\lambda = \begin{cases} m & \text{if } \lambda = \mu \\ 0 & \text{if } \lambda \neq \mu \end{cases}$ (cf. 1.3.19) (so that $(m_\lambda)_{\lambda\in\Lambda} \in \prod_{\lambda\in\Lambda} M_\lambda$ is an element of the direct sum iff for all but finitely many of the $\lambda \in \Lambda$, $m_\lambda = 0$). Each of these constructions enjoys a universal property. For any R-module N there are bijections:

$$(*48)\quad \begin{cases} \mathrm{Hom}(N, \Pi_{\lambda\in\Lambda}M_\lambda) \xrightarrow{\sim} \Pi_{\lambda\in\Lambda}\mathrm{Hom}(N, M_\lambda) : f \mapsto (p_\lambda \circ f)_{\lambda\in\Lambda} \\ \mathrm{Hom}(\oplus_{\lambda\in\Lambda}M_\lambda, N) \xrightarrow{\sim} \Pi_{\lambda\in\Lambda}\mathrm{Hom}(M_\lambda, N) : f \mapsto (f \circ i_\lambda)_{\lambda\in\Lambda} \end{cases}$$

where $p_\mu : \Pi_{\lambda\in\Lambda}M_\lambda \to M_\mu$ is the μth projection.

If now $(X, \mathcal{O})$ is a ringed space over R and $(M_\lambda)_{\lambda\in\Lambda}$ is a family of $\mathcal{O}$-Modules, it is easily verified that the following presheaves are sheaves:

$$\left.\begin{array}{l} U \mapsto \Pi_{\lambda\in\Lambda}\Gamma(U, M_\lambda) \\ U \mapsto \oplus_{\lambda\in\Lambda}\Gamma(U, M_\lambda) \end{array}\right\} \quad \text{for } U \text{ open in } X$$

and so define $\mathcal{O}$-Modules called the <u>direct product</u> $\Pi_{\lambda\in\Lambda}M_\lambda$ and the <u>direct sum</u> $\oplus_{\lambda\in\Lambda}M_\lambda$ of the family. There are natural projection morphisms $p_\mu : \Pi_{\lambda\in\Lambda}M_\lambda \to M_\mu$ and injection morphisms $i_\mu : M_\mu \to \oplus_{\lambda\in\Lambda}M_\lambda$, and the universal properties (*48) hold as stated, in the category $\mathcal{O}$-Mod. Hence the abelian category $\mathcal{O}$-Mod satisfies Grothendieck's axioms AB3 and AB3* (existence of infinite sums and products) ([T] §1. 5). Of course, the case $\mathcal{O} = \mathbf{Z}_X$, Λ having two elements was dealt with in §3. 5.

Since direct sums and stalks are both defined as colimits, we have for $x \in X$ the stalk:

$$\begin{aligned}(\oplus_{\lambda\in\Lambda}M_\lambda)_x &= \varinjlim_{U\ni x}(\oplus_{\lambda\in\Lambda}\Gamma(U, M_\lambda)) \\ &= \oplus_{\lambda\in\Lambda}\varinjlim_{U\ni x}\Gamma(U, M_\lambda) \\ &= \oplus_{\lambda\in\Lambda}(M_\lambda)_x\end{aligned}$$

as can easily be verified directly. However there is no corresponding expression for the stalks of the direct product.

4. 9 Construction. Let $(X, \mathcal{O})$ be a ringed space over a ring R. If L, M are $\mathcal{O}$-Modules we can define a presheaf on X by

$$U \to \Gamma(U, L) \otimes_{\Gamma(U, \mathcal{O})}\Gamma(U, M)$$

for U open in X (tensor product of $\Gamma(U, \mathcal{O})$-modules). The sheafification of this presheaf is an $\mathcal{O}$-Module called the <u>tensor product</u> of L and M over $\mathcal{O}$ and denoted by $L \otimes_{\mathcal{O}} M$. See 5. 8 for an example where the presheaf is not a sheaf.

Since sheafification does not change the stalks (2. 4. 5), and tensor product commutes with direct limits (see for example Bourbaki, Algebra, Ch. II §6. 7), $L \otimes_{\mathcal{O}} M$ has stalk

$$(L \otimes_{\mathcal{O}} M)_x = \varinjlim_{U\ni x}\Gamma(U, L) \otimes_{\Gamma(U, \mathcal{O})}\Gamma(U, M) = L_x \otimes_{\mathcal{O}_x} M_x$$

at $x \in X$; also if s, t are sections of L, M respectively over an open set U then the map

$$x \mapsto s_x \otimes t_x \quad \text{for } x \in U$$

is a section of $L \otimes_{\mathcal{O}} M$ over U.

It is easy to see from the definition that for any $\mathcal{O}$-Module M we have $M \otimes_{\mathcal{O}} \mathcal{O} \cong M$.

4.10 Tensor product of Modules inherits many of the properties of tensor product of modules; for instance, tensor product with a fixed $\mathcal{O}$-Module M gives a right exact covariant functor

$$M \otimes_{\mathcal{O}} - : \mathcal{O}\text{-Mod} \to \mathcal{O}\text{-Mod}.$$

There is also a universal property as follows. If L, M, N are $\mathcal{O}$-Modules, a <u>bilinear map</u> from L, M to N is a map of sheaves of sets $f : L \oplus M \to N$ such that for each open set U the map

$$\Gamma(U, f) : \Gamma(U, L) \oplus \Gamma(U, M) \to \Gamma(U, N)$$

is a bilinear map of $\Gamma(U, \mathcal{O})$-modules. Such maps form a set $\mathrm{Bilin}(L, M; N)$ which is covariant-functorial in N.

Proposition. $L \otimes_{\mathcal{O}} M$ <u>represents the functor</u> $\mathrm{Bilin}(L, M; -) : \mathcal{O}\text{-Mod} \to \mathrm{Sets}$; <u>that is, there is an isomorphism</u>

$$\mathrm{Bilin}(L, M; N) \xrightarrow{\sim} \mathrm{Hom}_{\mathcal{O}}(L \otimes_{\mathcal{O}} M, N)$$

<u>which is natural in the</u> $\mathcal{O}$-<u>Module</u> N. <u>This property characterises the</u> $\mathcal{O}$-<u>Module</u> $L \otimes_{\mathcal{O}} M$ <u>up to isomorphism.</u>

Proof. Direct from the corresponding proposition for modules over a ring, and the universal property of sheafification (2.4.2). //

4.11 Suppose that $(X, \mathcal{O})$ is a ringed space over a ring R, and U is open in X. Then U is a ringed space with structure sheaf $\mathcal{O} | U$ (3.8.1 and 1.4), and if M is an $\mathcal{O}$-Module, it is easy to see (using 3.8.3) that $M | U$ is in a natural way an $(\mathcal{O} | U)$-Module.

Proposition. If L, M are $\mathcal{O}$-Modules and U is open in X, then there is a natural isomorphism of $(\mathcal{O}|U)$-Modules

$$(L|U) \otimes_{\mathcal{O}|U} (M|U) \xrightarrow{\sim} (L \otimes_{\mathcal{O}} M)|U.$$

Proof. There is a natural bilinear map

$$(L|U) \oplus (M|U) \to (L \otimes_{\mathcal{O}} M)|U$$

obtained by restricting $L \oplus M \to L \otimes_{\mathcal{O}} M$ to U. By 4.10 this induces the required map, which is an isomorphism since on the stalk at $x \in U$ it reduces, after the identification of 4.9, to the identity map

$$L_x \otimes M_x \to L_x \otimes M_x$$

(using 3.7.12 or 3.8.3). //

4.12 **Construction.** Let $(X, \mathcal{O})$ be a ringed space over a ring R. If M, N are $\mathcal{O}$-Modules, define the presheaf $\underline{\mathrm{Hom}}(M, N)$ by

$$\underline{\mathrm{Hom}}(M, N)(U) = \mathrm{Hom}_{\mathcal{O}|U}(M|U, N|U) \text{ for } U \text{ open in } X,$$

where the right-hand side is the set of $(\mathcal{O}|U)$-Module morphisms of $M|U$ into $N|U$, and has a natural structure of $\Gamma(U, \mathcal{O})$-module. If $U \supseteq V$ are open, we have a map

$$\mathrm{Hom}_{\mathcal{O}|U}(M|U, N|U) \xrightarrow{\alpha^*} \mathrm{Hom}_{\mathcal{O}|V}(M|V, N|V)$$

where $\alpha : V \hookrightarrow U$ is the inclusion (3.7.11), and α^* is easily seen to be a module morphism over the restriction $\Gamma(U, \mathcal{O}) \to \Gamma(V, \mathcal{O})$. If we reinterpret the elements of $\mathrm{Hom}_{\mathcal{O}|U}(M|U, N|U)$ as continuous maps between certain subspaces of the sheaf spaces LM and LN (cf. 3.8.3) we can check readily that $\underline{\mathrm{Hom}}(M, N)$ is a sheaf of $\mathcal{O}$-modules, called the sheaf of germs of homomorphisms from M to N, or sheaf-hom for short; if we wish to emphasise the structure sheaf we can write $\underline{\mathrm{Hom}}_{\mathcal{O}}(M, N)$.

4.13 Suppose that $\Phi : (X, \mathcal{O}_X) \to (Y, \mathcal{O}_Y)$ is a morphism of ringed spaces over a ring R, with underlying continuous map $\phi : X \to Y$ and morphism of sheaves of R-algebras $\psi : \mathcal{O}_Y \to \phi_* \mathcal{O}_X$.

If M is an $\mathcal{O}_X$-Module, the definition

$$\Gamma(V, \phi_* M) = \Gamma(\phi^{-1} V, M) \text{ for } V \text{ open in } Y$$

shows that $\phi_* M$ is naturally a $(\phi_* \mathcal{O}_X)$-Module (cf. 1.4). The morphism ψ gives for each open V in Y a change-of-rings morphism $\Gamma(V, \mathcal{O}_Y) \to \Gamma(V, \phi_* \mathcal{O}_X)$ which enables us to regard $\phi_* M$ as an $\mathcal{O}_Y$-Module. Hence Φ induces a functor

$$\Phi_* : \mathcal{O}_X\text{-Mod} \to \mathcal{O}_Y\text{-Mod}.$$

If N is an $\mathcal{O}_Y$-Module, then $\phi^* N$ is a sheaf of abelian groups on X (3.7.11), and an easy examination of its groups of sections shows that $\phi^* N$ has a natural structure of $(\phi^* \mathcal{O}_Y)$-Module. We also have a change-of-rings morphism $\phi^* \mathcal{O}_Y \to \mathcal{O}_X$ deduced from ψ by 3.7.13, and we define

$$\Phi^* N = \phi^* N \otimes_{\phi^* \mathcal{O}_Y} \mathcal{O}_X$$

to get a functor

$$\Phi^* : \mathcal{O}_Y\text{-Mod} \to \mathcal{O}_X\text{-Mod}.$$

4.14 **Theorem.** _If_ $\Phi : (X, \mathcal{O}_X) \to (Y, \mathcal{O}_Y)$ _is a morphism of ringed spaces over a ring_ R, _then_ Φ^* _is left adjoint to_ Φ_*; _that is there is a natural bijection_

$$\mathrm{Hom}_{\mathcal{O}_X}(\Phi^* N, M) \xrightarrow{\sim} \mathrm{Hom}_{\mathcal{O}_Y}(N, \Phi_* M)$$

whenever $M \in \mathrm{Ob}\, \mathcal{O}_X\text{-Mod}$ _and_ $N \in \mathrm{Ob}\, \mathcal{O}_Y\text{-Mod}$.

Proof. Let $\phi : X \to Y$ be the underlying continuous map of Φ. Then we can construct the following diagram:

$$\begin{array}{ccc}
\mathrm{Hom}_{Z_X}(\phi^*N,\ M) & \xrightarrow[f]{\sim} & \mathrm{Hom}_{Z_Y}(N,\ \phi_*M) \\
\uparrow & & \uparrow \\
\mathrm{Hom}_{\phi^*\mathcal{O}_Y}(\phi^*N,\ M) & \xrightarrow[g]{\sim} & \mathrm{Hom}_{\mathcal{O}_Y}(N,\ \Phi_*M) \\
\wr\uparrow h & & \\
\mathrm{Hom}_{\mathcal{O}_X}(\phi^*N \otimes_{\phi^*\mathcal{O}_Y} \mathcal{O}_X,\ M) & &
\end{array}$$

where the bijection f is given by 3.7.13, and is easily seen to induce the bijection g, while the map h is given by composition with the morphism

$$\phi^*N \to \phi^*N \otimes_{\phi^*\mathcal{O}_Y} \mathcal{O}_X$$

(given on sections by $s \mapsto s \otimes 1$) and has inverse sending a morphism $u : \phi^*N \to M$ to the composite

$$\phi^*N \otimes_{\phi^*\mathcal{O}_Y} \mathcal{O}_X \xrightarrow{u \otimes id} M \otimes_{\phi^*\mathcal{O}_Y} \mathcal{O}_X \xrightarrow{v} M,$$

where the multiplication morphism v is given on sections by $s \otimes t \mapsto ts$ (since M is an $\mathcal{O}_X$-Module). //

4.15 If $(X, \mathcal{O}_X)$ is a ringed space over a ring R and U is an open subspace of X, then $\mathcal{O}_X | U$ (3.8.1) is a sheaf of R-algebras on U which makes U into a ringed space over R and gives a morphism of ringed spaces

$$\Phi : (U,\ \mathcal{O}_X | U) \to (X,\ \mathcal{O}_X)$$

over the inclusion map $U \hookrightarrow X$ (the map of sheaves is given by the natural map of 3.7.11(i)). If M is an $\mathcal{O}_X$-Module, then Φ^*M (as in 4.14) is the $(\mathcal{O}_X | U)$-Module $M | U$.

4.5 Locally free Modules

5.1 **Definition.** If $(X, \mathcal{O})$ is a ringed space over a ring R, an $\mathcal{O}$-Module M is called locally free of rank $n \in \mathbf{N}$ iff X can be covered by open sets U such that

$$M|U \cong (\mathcal{O}|U)^n \text{ as } (\mathcal{O}|U)\text{-Modules}$$

where for a Module N, $N^n = \oplus_{i=1}^n N$ is the biproduct of n copies of N.

5.2 If $(X, \mathcal{O})$ is a ringed space over a ring R, the endomorphism ring of $\mathcal{O}$

$$\mathrm{End}(\mathcal{O}) = \mathrm{Hom}_{\mathcal{O}}(\mathcal{O}, \mathcal{O})$$

is the set of $\mathcal{O}$-Module morphisms: $\mathcal{O} \to \mathcal{O}$, with composition as multiplication. Its group of units is the group $\mathrm{Aut}(\mathcal{O})$ of $\mathcal{O}$-Module automorphisms of $\mathcal{O}$. We have a map

$$\Gamma(X, \mathcal{O}) \to \mathrm{End}(\mathcal{O})$$

sending $s \in \Gamma(X, \mathcal{O})$ to the endomorphism which is given over an open U by multiplication by the restriction $\rho_U^X(s)$:

$$\Gamma(U, \mathcal{O}) \to \Gamma(U, \mathcal{O}) : t \mapsto \rho_U^X(s).t.$$

5.3 **Proposition.** In the situation of 5.2, the map

$$\Gamma(X, \mathcal{O}) \to \mathrm{End}(\mathcal{O})$$

is a ring isomorphism. Hence $\mathrm{Aut}(\mathcal{O}) \xrightarrow{\sim} \Gamma(X, \mathcal{O})^*$ (the group of units of the ring $\Gamma(X, \mathcal{O})$) and for $n \in \mathbf{N}$

$$\mathrm{End}(\mathcal{O}^n) \xrightarrow{\sim} M_n(\Gamma(X, \mathcal{O})) = \mathrm{End}_{\Gamma(X, \mathcal{O})}(\Gamma(X, \mathcal{O})^n)$$

can be identified with the ring of $n \times n$ matrices over $\Gamma(X, \mathcal{O})$, while

$$\mathrm{Aut}(\mathcal{O}^n) \xrightarrow{\sim} GL_n(\Gamma(X, \mathcal{O})) = \mathrm{Aut}_{\Gamma(X, \mathcal{O})}(\Gamma(X, \mathcal{O})^n).$$

Proof. The map of 5.2 is clearly a ring morphism, and it is injective, since if $s \in \Gamma(X, \mathcal{O})$ gives the zero morphism $\mathcal{O} \to \mathcal{O}$, then in particular

$$s = \rho_X^X(s).1 = 0.$$

Given $f \in \mathrm{End}(\mathcal{O})$, each $\Gamma(U, f) : \Gamma(U, \mathcal{O}) \to \Gamma(U, \mathcal{O})$ is a $\Gamma(U, \mathcal{O})$-module morphism, and so is multiplication by some $t_U \in \Gamma(U, \mathcal{O})$ (namely $t_U = \Gamma(U, f)(1)$); then the commutativity of the square

$$\begin{array}{ccc} \Gamma(X, \mathcal{O}) & \longrightarrow & \Gamma(X, \mathcal{O}) \\ \downarrow & & \downarrow \\ \Gamma(U, \mathcal{O}) & \longrightarrow & \Gamma(U, \mathcal{O}) \end{array}$$

shows that $t_U = \rho_U^X(t_X)$. Hence $f \mapsto t_X = \Gamma(U, f)(1)$ gives an inverse to the morphism of 5.2.

If multiplication by $s \in \Gamma(X, \mathcal{O})$ is an automorphism of $\mathcal{O}$, then $t \mapsto s.t$ is surjective on $\Gamma(X, \mathcal{O})$, so s is a unit; the converse is clear.

The last result follows from the usual procedure of considering the various composites

$$\mathcal{O} \to \mathcal{O}^n \to \mathcal{O}^n \to \mathcal{O}$$

to obtain the matrix components. //

5.4 Examples

A. If X is a manifold of the kind described in 3.7A, B, C with structure sheaf $\mathcal{O}$, then the locally free $\mathcal{O}$-Modules of rank n are just the sheaves of sections of rank n vector bundles. For it is easy to see that such a sheaf is locally free, by considering its restriction to the open sets of a trivialisation. Conversely, given a locally free Module M, we can patch together copies of R^n over the intersection $U \cap V$ of two open sets over which M is free by means of the isomorphism

$$(\mathcal{O}|U \cap V)^n \cong (\mathcal{O}|U)^n|V \cong M|U \cap V \cong (\mathcal{O}|V)^n|U \cong (\mathcal{O}|U \cap V)^n$$

which by 5.3 is given by a member of the general linear group $GL_n(\Gamma(U \cap V, \mathcal{O}))$ (that is, an invertible $n \times n$ matrix of R-valued functions on $U \cap V$).

B. It can be shown that the locally free Modules over an affine scheme $\mathrm{Spec}\, R$ are the Modules of the form $\tilde{P}$ (4.5B) where P is a projective R-module (of constant finite rank). (References: Bourbaki, Algèbre Commutative, Ch. II, §§5.2 and 5.3, especially Th. 2; [K], 3.3.7.)

5.5 **Definition.** The locally free Modules of rank 1 over a ringed space $(X, \mathcal{O})$ are called the invertible $\mathcal{O}$-Modules. We shall see later (5.4.16) (or we could check directly) that there is only a set of isomorphism classes of invertible $\mathcal{O}$-Modules; taking this on trust, this set is called the picard group $\operatorname{Pic} X$ of the ringed space $(X, \mathcal{O})$. The justification for these terms is given by the following result.

5.6 **Theorem.** *If $(X, \mathcal{O})$ is a ringed space over a ring R, then $\operatorname{Pic} X$ is a group under tensor product, with identity the isomorphism class of the free module $\mathcal{O}$, and with the inverse of the class of an invertible sheaf M being given by the class of $\underline{\operatorname{Hom}}(M, \mathcal{O})$ (the 'dual' of M).*

Proof. We saw in 4.9 that for any $\mathcal{O}$-Module M, $M \otimes_{\mathcal{O}} \mathcal{O} \cong M$. If M, N are locally free of rank 1, then for all members U of a sufficiently fine open cover we have (after 4.11)

$$(M \otimes_{\mathcal{O}} N)|U \cong M|U \otimes_{\mathcal{O}|U} N|U \cong \mathcal{O}|U \otimes_{\mathcal{O}|U} \mathcal{O}|U \cong \mathcal{O}|U$$

so that $M \otimes_{\mathcal{O}} N$ is also invertible.

We must show that there is an isomorphism

$$f : \underline{\operatorname{Hom}}(M, \mathcal{O}) \otimes_{\mathcal{O}} M \to \mathcal{O}.$$

We have a map from the presheaf defining the LHS to $\mathcal{O}$ by

$$\operatorname{Hom}_{\mathcal{O}|U}(M|U, \mathcal{O}|U) \otimes_{\Gamma(U, \mathcal{O})} \Gamma(U, M) \to \Gamma(U, \mathcal{O}) : g \otimes s \mapsto \Gamma(U, g)(s)$$

and so a morphism f exists by 2.4.2. But if U is such that M is trivial (free) over U, using $M|U \cong \mathcal{O}|U$ we have the isomorphism

$$\operatorname{Hom}_{\mathcal{O}|U}(\mathcal{O}|U, \mathcal{O}|U) \xrightarrow{\sim} \Gamma(U, \mathcal{O})$$

of 5.3; hence $\forall x \in X$ the stalk morphism f_x is an isomorphism, and so by 3.4.10 f is an isomorphism. //

5.7 **Exercise** (for Commutative Algebraists and Number Theorists). Show that if A is an integral ring, $\operatorname{Pic}(\operatorname{Spec} A)$ is the group of classes of fractional ideals of A; in particular, if A is the ring of integers of a number field,

$$\mathrm{Pic}(\mathrm{Spec}\ A) = \text{the ideal class group of } A$$

and so is finite.

5.8 **Example.** Let U, V be two copies of **C** with coordinate functions $u : U \to \mathbf{C}$ and $v : V \to \mathbf{C}$. Then $U_1 = \{x \in U;\ u(x) \neq 0\}$ and $V_1 = \{x \in V;\ v(x) \neq 0\}$ are isomorphic by the recipe $uv = 1$; that is, by the map

$$\phi : U_1 \to V_1 : x \mapsto y \text{ where } v(y) = \frac{1}{u(x)}\ .$$

Glueing U, V together by ϕ we obtain a space X with open subsets U', V' and maps

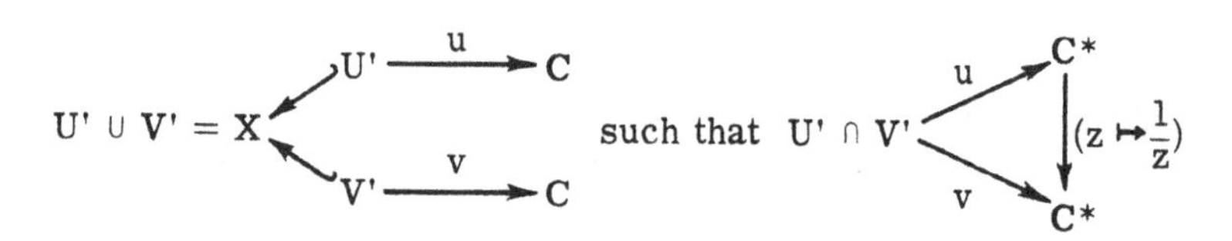

commutes. This determines the topology on X. In fact X is $\mathbf{P}^1(\mathbf{C})$ (the Riemann sphere) with homogeneous coordinates

$$(u,\ 1) = (u,\ uv) = (1,\ v)$$

(at any point one of these expressions is well-defined).

The map ϕ is an isomorphism for any reasonable structure on U, V. Let us drop the distinction between U and U', and V and V', and for the sake of definiteness let $\mathcal{O}_U$, $\mathcal{O}_V$ be the sheaves of **C**-valued analytic functions on U, V (so that $(U, \mathcal{O}_U) \cong (\mathbf{C}, \mathbf{C}^\omega) \cong (V, \mathcal{O}_V)$). Then these define a unique sheaf of rings $\mathcal{O}_X$ on X giving X the structure of **C**-analytic manifold. Indeed, for W open in X

$$\Gamma(W, \mathcal{O}_X) \cong \{(f, g) \in \Gamma(W \cap U, \mathcal{O}_U) \times \Gamma(W \cap V, \mathcal{O}_V);$$
$$\forall x \in U \cap V \quad f(u(x)) = g(v(x))\}.$$

By Liouville's theorem, $\Gamma(X, \mathcal{O}_X) \cong \mathbf{C}$ consists of just the constant functions.

For $n \in \mathbf{Z}$ we can define an invertible $\mathcal{O}_X$-Module $\mathcal{O}_X(n)$ on X as follows. For W open in X, let

$$\Gamma(W, \mathcal{O}_X(n)) = \{(f, g) \in \Gamma(W \cap U, \mathcal{O}_U) \times \Gamma(W \cap V, \mathcal{O}_V);$$
$$\forall x \in U \cap V \quad f(u(x)) = u(x)^n g(v(x))\}.$$

In other words, $\mathcal{O}_X(n)$ is obtained by glueing together $\mathcal{O}_X|U = \mathcal{O}_U$ and $\mathcal{O}_X|V = \mathcal{O}_V$ by means of the isomorphism over $U \cap V$

$$\begin{array}{ccc} \mathcal{O}_X|U \cap V & & \mathcal{O}_X|V \cap U \\ \wr\| & & \wr\| \\ \mathcal{O}_C|C^* & \xrightarrow[\text{by } z^n]{\text{multiply}} & \mathcal{O}_C|C^* \end{array}$$

where $C^* = \{z \in \mathbf{C};\ z \neq 0\}$.

Then we have

$$\Gamma(X, \mathcal{O}_X(n)) \cong \begin{cases} 0 & \text{for } n < 0 \\ \text{(abelian group of homogeneous polynomials of degree } n \text{ in } x, y) & \text{for } n \geq 0 \end{cases}$$

(where notionally $u=x/y$ and $v=y/x$), and it is easy to check that

$$\mathcal{O}_X(n) \otimes_{\mathcal{O}_X} \mathcal{O}_X(m) \cong \mathcal{O}_X(n+m)$$
$$\underline{\mathrm{Hom}}(\mathcal{O}_X(n), \mathcal{O}_X) \cong \mathcal{O}_X(-n)$$

so that $\mathcal{O}_X(n) = (\mathcal{O}_X(1))^n$ in $\operatorname{Pic} X$. Taking for example $n = m = 1$ we see that $\Gamma(X, \mathcal{O}_X(1)) \otimes_{\Gamma(X, \mathcal{O}_X)} \Gamma(X, \mathcal{O}_X(1)) \not\cong \Gamma(X, \mathcal{O}_X(2))$ (they are $\mathbf{C}$-vectorspaces of dimensions $2.2 = 4$ and 3); this provides the example advertised in 4.9.

We have shown that for $X = \mathbf{P}^1(\mathbf{C})$, $\operatorname{Pic} X$ has an infinite cyclic subgroup generated by $\mathcal{O}_X(1)$; it can be shown that this is all of $\operatorname{Pic} X$.

Exercises on Chapter 4

1. Let X be a topological space and R be a ring. The category of sheaves of R-modules (compare 3. Ex. 8) on X is (isomorphic to) the

category of $\mathcal{O}$-Modules, where $\mathcal{O}$ is the constant sheaf R.

(a) If L, M are two sheaves of R-modules on X, then $L \otimes_R M$ denotes the sheafification of the presheaf $U \mapsto \Gamma(U, L) \otimes_R \Gamma(U, M)$ (U open in X), in conformity with 4.9.

(b) If Y is another topological space, and N a sheaf of R-modules on Y, show that there is a unique sheaf $L \hat{\otimes}_R N$ (called the <u>total tensor product</u>) on the product space $X \times Y$ with the properties

(i) $\forall (x, y) \in X \times Y \quad (L \hat{\otimes}_R N)_{(x, y)} \cong L_x \otimes_R N_y$

(ii) if U is open in X and V is open in Y, and $s \in \Gamma(U, L)$ and $t \in \Gamma(V, N)$, then

$$(x, y) \mapsto s_x \otimes t_y \in L_x \otimes_R N_y \cong (L \hat{\otimes}_R N)_{(x, y)}$$

is a section of $L \hat{\otimes}_R N$ over the open set $U \times V$.

(c) Show that $L \hat{\otimes}_R N = (\pi_1^* L) \otimes_R (\pi_2^* N)$, where $\pi_1 : X \times Y \to X$ and $\pi_2 : X \times Y \to Y$ are the projections.

(d) Show that $L \otimes_R M = d^*(L \hat{\otimes}_R M)$ where $d : X \to X \times X$ is the diagonal map.

(e) Discover the universal property satisfied by the total tensor product construction. (Compare [G] 2.10.)

2. Show that we can remove the condition in 2.6 that the basis $\mathcal{U}$ be closed under finite intersections, provided we replace the condition on the 'presheaf F defined only on the basis $\mathcal{U}$' by the following condition:

whenever $U = \cup_{\lambda \in \Lambda} U_\lambda$ with $U \in \mathcal{U}$ and $\forall \lambda \in \Lambda \; U_\lambda \in \mathcal{U}$,

for any set (or abelian group or ring as appropriate) T, the map

$$\begin{aligned} \mathrm{Hom}(T, F(U)) &\to \Pi_{\lambda \in \Lambda} \mathrm{Hom}(T, F(U_\lambda)) \\ f &\mapsto (\rho^U_{U_\lambda} \circ f)_{\lambda \in \Lambda} \end{aligned}$$

is an injection, with image

$$\{(f_\lambda)_{\lambda \in \Lambda} ; \forall \lambda, \mu \in \Lambda, \; \forall V \in \mathcal{U} \text{ such that } V \subseteq U_\lambda \cap U_\mu \;\; \rho_V^{U_\lambda} \circ f_\lambda = \rho_V^{U_\mu} \circ f_\mu \}.$$

Show that, even in this more general setting, the alternative construction indicated in 2.6 (setting $\Gamma(V, G) = \varprojlim \Gamma(U, G)$) works. (Compare [EGAI] 0, 3.2 and 3. Ex. 8.)

3. Draw a picture of Spec $\mathbf{Z}[t]$. (Compare Mumford, Introduction to Algebraic Geometry, p. 141.)

4. Show that for any ring R, Spec R is compact.

5. Let R be a ring and $f \in R$. Show that the natural ring morphism $R \to R_f = R[\frac{1}{f}] = S^{-1}R$ with $S = \{f^n;\ n \in \mathbf{N}\}$ induces an isomorphism of ringed spaces between

$$(\text{Spec } R_f,\ \mathcal{O}_{\text{Spec } R_f}) \quad \text{and} \quad (D(f),\ \phi^*\mathcal{O}_{\text{Spec } R} = \mathcal{O}_{\text{Spec } R}|D(f))$$

where $\phi : D(f) \hookrightarrow \text{Spec } R$ is the inclusion of the open subspace $D(f)$.

6. Let R be a ring and $\mathfrak{a}$ an ideal of R. Show that the natural ring morphism $R \to R/\mathfrak{a}$ induces a morphism of ringed spaces $\text{Spec}(R/\mathfrak{a}) \to \text{Spec } R$ which is a homeomorphism of $\text{Spec}(R/\mathfrak{a})$ with the closed subspace $V(\mathfrak{a})$ of Spec R; but show by example that it need not induce an isomorphism of ringed spaces between $\text{Spec}(R/\mathfrak{a})$ and

$$(V(\mathfrak{a}),\ \mathcal{O}_{\text{Spec } R}|V(\mathfrak{a})).$$

7. For $k \in \mathbf{N}$, let $S^k = \{x \in \mathbf{R}^{k+1};\ \|x\| = 1\}$. Letting $\mathcal{O}$ be the sheaf of continuous (or differentiable, or analytic) $\mathbf{R}$-valued functions on $\mathbf{R}^{k+1}$, define a sheaf of ideals I in $\mathcal{O}|S^k = \phi^*\mathcal{O}$ (where $\phi : S^k \hookrightarrow \mathbf{R}^{k+1}$) with stalks

$$I_y = \{f \in (\mathcal{O}|S^k)_y;\ f \text{ is the germ of a function } f' \text{ such that } f'|S^k = 0\}.$$

and show that putting $\mathcal{O}'$ to be the quotient of $\mathcal{O}|S^k$ by I, the ringed space $(S^k, \mathcal{O}')$ is a k-dimensional topological (or differentiable, or analytic) manifold, the <u>k-sphere.</u>

8. (Projective n-space)

(a) (Topological version) Let $K = \mathbf{R}$ or $\mathbf{C}$, and let V be an $(n+1)$-dimensional vectorspace over K. Let $P = \mathbf{P}^n_k$ be the set of hyperplanes (= n-dimensional subspaces) in V. Show that P is a quotient of a sphere S^k of dimension $k = n$ (if $K = \mathbf{R}$) or $2n+1$ (if $K = \mathbf{C}$) by an equivalence relation whose equivalence classes have two elements if $K = \mathbf{R}$, or biject with S^1 if $X = \mathbf{C}$. Giving P the quotient topology show that P has a natural structure of analytic (or differentiable, or topological) manifold.

[Hint: the map $S^k \to P$ is a covering map; cover S^k by hemispheres and use these to give P a structure sheaf.]

(b) (Algebraic version) Let K be any field. Use the ring isomorphisms

$$K[\frac{X_0}{X_i}, \frac{X_1}{X_i}, \ldots, \frac{X_n}{X_i}][\frac{1}{X_j}] \cong K[\frac{X_0}{X_j}, \frac{X_1}{X_j}, \ldots, \frac{X_n}{X_j}][\frac{1}{X_i}] \text{ for } 0 \le i, j \le n$$

to glue together the $n+1$ copies $U_i = \operatorname{Spec} R_i$ of $\mathbf{A}^n_K$, where

$$R_i = K[\frac{X_0}{X_i}, \ldots, \frac{X_n}{X_i}] \cong K[Y_1, \ldots, Y_n],$$

into a scheme $\mathbf{P}^n_K$.

Show that if K is algebraically closed, the closed points of $\mathbf{P}^n_K$ biject with the (n+1)-tuples $(\alpha_0, \ldots, \alpha_n) \in K^{n+1}$ considered modulo the relation

$$(\alpha_0, \ldots, \alpha_n) \sim (\lambda\alpha_0, \ldots, \lambda\alpha_n) \text{ for } \lambda \in K \setminus \{0\}.$$

(Homogeneous coordinates.) Show that $\Gamma(\mathbf{P}^n_K, \mathcal{O}) \cong K$.

Working by analogy with 5.8, construct on $\mathbf{P}^n_K$ an invertible sheaf $\mathcal{O}(1)$ with the property that

$$\Gamma(\mathbf{P}^n_K, \mathcal{O}(1)) \cong \text{the set of homogeneous linear forms in } X_0, \ldots, X_n.$$

If you are ambitious, show that the points of $\mathbf{P}^n_K$ biject naturally with the homogeneous prime ideals of the graded ring $K[X_0, \ldots, X_n]$ which do not contain the ideal $(X_0, \ldots, X_n)$. Deduce an analogous construction of a scheme Proj(S) for any graded ring S. (Reference [EGA II], 2.3.)

9. (Algebraic curves)

(a) (Topological version) Show that the subset of $\mathbf{R}^2$ (or of $\mathbf{C}^2$ if you are ambitious) given by

$$C = \{(x, y);\ y^2 = (x + 1)x(x - 1)\}$$

is a manifold (topological, differentiable or analytic as you wish), whereas that given by the equation $y^2 = x^2(x + 1)$ is not. (Draw a picture.)

(b) (Algebraic version) Let K be any field and $f(x, y) \in K[x, y] = R$. By Q6 the morphism $R \to R/Rf$ identifies the space $\operatorname{Spec}(R/Rf)$ with a closed subspace C of $\operatorname{Spec} R = \mathbf{A}^2_{\mathbf{K}} = X$ say. Find the sheaf of ideals I in $\mathcal{O}_{\mathbf{X}}|C$ such that, letting $\mathcal{O}'$ be the quotient of $\mathcal{O}_{\mathbf{X}}|C$ by I, the morphism

$$\operatorname{Spec}(R/Rf) \to (C, \mathcal{O}')$$

is an isomorphism of ringed spaces (and so of affine schemes). (Compare Q7.)

Putting $f(x, y) = y^2 - x^3 + x$ or $y^2 - x^3 - x^2$ we get the algebraic analogues of the curves of part (a). What is different about the two cases?

Other curves for your amusement:

$$y^2 = x^2(x - 1)$$
$$y^2 = x^3.$$

10. Show that the construction of the prime spectrum of a ring can be generalised as follows. Let X be a scheme, with structure sheaf $\mathcal{O}$. Let A be a sheaf of $\mathcal{O}$-algebras (an $\mathcal{O}$-<u>Algebra</u>), that is an $\mathcal{O}$-Module such that for each U open in X, the $\Gamma(U, \mathcal{O})$-module $\Gamma(U, A)$ has a structure of $\Gamma(U, \mathcal{O})$-algebra, in a way compatible with restriction maps (aliter, there is a given multiplication $A \otimes_{\mathcal{O}} A \to A$ and a section $1 \in \Gamma(X, A)$ satisfying the usual laws for an algebra). Suppose that A is quasi-coherent as an $\mathcal{O}$-Module.

For each open affine $U = \operatorname{Spec} R$ of X, $A|U$ is an $\mathcal{O}|U$-Algebra and so we get a scheme morphism

$$\operatorname{Spec} \Gamma(U, A) \to \operatorname{Spec} R = U;$$

for different U, these fit together into a scheme morphism

$$(*)\qquad \underline{\mathrm{Spec}}\ A \to X$$

where $\underline{\mathrm{Spec}}\ A$ is obtained by glueing the different $\mathrm{Spec}\ \Gamma(U, A)$ by means of the isomorphisms provided by the structure of A.

(a) If $X = \mathrm{Spec}\ R$ is affine, then so is $\underline{\mathrm{Spec}}\ A$; indeed it is the spectrum of an R-algebra. For this reason, morphisms of the type (*) are called affine morphisms.

(b) Show that a scheme morphism $f : Y \to X$ is affine iff X can be covered by affine open sets U such that each $f^{-1}(U)$ is affine. (Hint: put $A = f_*(\mathcal{O}_Y)$; see [EGA I] §9.1.)

(c) For example, for any X let $A = \mathcal{O}[t]$ be defined by having sections

$$\Gamma(U, A) = \Gamma(U, \mathcal{O})[t] \qquad \text{for } U \text{ open in } X$$

(the polynomial ring). Then $\underline{\mathrm{Spec}}\ A = \underline{\mathrm{Spec}}\ \mathcal{O}[t]$ is called the affine line $\mathbf{A}^1_X$ over X. For $X = \mathrm{Spec}\ R$, $\mathbf{A}^1_X = \mathrm{Spec}\ R[t]$. Show that for any scheme X, $\mathbf{A}^1_X$ is the pullback in the category of schemes of the diagram

$$\begin{array}{ccc} & & \mathbf{A}^1_{\mathbf{Z}} = \mathrm{Spec}\ \mathbf{Z}[t] \\ & & \downarrow \\ X & \longrightarrow & \mathrm{Spec}\ \mathbf{Z} \end{array}$$

(that is, the product $X \times \mathbf{A}^1_{\mathbf{Z}}$, since $\mathrm{Spec}\ \mathbf{Z}$ is final in the category of schemes).

(d) Generalising (c), show that any $\mathcal{O}$-Module E gives rise to an $\mathcal{O}$-Algebra SE constructed locally out of symmetric algebras, and hence to a scheme morphism

$$\mathbf{V}(E) = \underline{\mathrm{Spec}}\ SE \to X.$$

(The example of (c) comes by putting $E = \mathcal{O}$.) If E is locally free (of rank n) show that $\mathbf{V}(E)$ is a 'vector bundle' over X, in the sense that it is locally of the form $U \times \mathbf{A}^n$ (where $\mathbf{A}^n = \mathrm{Spec}\ \mathbf{Z}[t_1, \ldots, t_n]$). (Compare [EGA I] §9.4.)

11. Let $(X, \mathcal{O})$ be a ringed space and M an $\mathcal{O}$-Module. M is said to be <u>of finite type</u> iff each $x \in X$ has an open neighbourhood U such that for some $n \in \mathbf{N}$

$$(*) \quad \begin{cases} \text{there is an exact sequence of } (\mathcal{O}|U)\text{-Modules} \\ (\mathcal{O}|U)^n \to (M|U) \to 0 \end{cases}$$

(where $(\mathcal{O}|U)^n = \oplus_{i=1}^n (\mathcal{O}|U)$).

Show that (*) is equivalent to

$$(**) \quad \begin{cases} \text{there are sections } s_1, s_2, \ldots, s_n \in \Gamma(U, M) \text{ such that for} \\ \text{each } y \in U \text{ the } \mathcal{O}_y\text{-module } M_y \text{ is generated by the germs} \\ (s_1)_y, \ldots, (s_n)_y. \end{cases}$$

Suppose that M is of finite type, that U is open in X with $x \in U$, and that $t_1, t_2, \ldots, t_m \in \Gamma(U, M)$ are such that the germs $(t_1)_x, \ldots, (t_m)_x$ generate the $\mathcal{O}_x$-module M_x. Prove that there is an open set V with $x \in V \subseteq U$ and such that for all $y \in V$, the germs $(t_1)_y, \ldots, (t_m)_y$ generate the $\mathcal{O}_y$-module M_y. [Hint: Express the $(s_i)_x$ of (**) as an $\mathcal{O}_x$-linear combination of the $(t_j)_x$.]

Now suppose that k is a field and that $(X, \mathcal{O})$ is in fact a ringed space over k with the property that for all $x \in X$

$$\mathcal{O}_x = k.$$

Let M be an $\mathcal{O}$-Module of finite type. Deduce that for each $x \in X$, M_x is a k-vectorspace of finite dimension. Deduce also that the function

$$e : X \to \mathbf{N} : x \mapsto \dim_k M_x$$

is upper semi-continuous (that is, for each $n \in \mathbf{N}$, $\{x \in X;\ e(x) \geq n\}$ is closed in X). This function provides X with a 'stratification' into locally closed subspaces $\{x \in X;\ e(x) = n\}$ over which the stalks of M have constant dimension. What further properties of this stratification can you see? If in addition X is compact?

Use this to give a simple example of an $\mathcal{O}$-Module M which is not of finite type.

12. Let $(X, \mathcal{O})$ be a ringed space and M an $\mathcal{O}$-Module. M is said to be of <u>finite presentation</u> iff each $x \in X$ has an open neighbourhood U such that for some $m, n \in \mathbf{N}$ there is an exact sequence of $(\mathcal{O}|U)$-Modules

$$(\mathcal{O}|U)^m \to (\mathcal{O}|U)^n \to (M|U) \to 0$$

(so that M is of finite type, and so are suitably chosen 'sheaves of relations' $\mathrm{Ker}((\mathcal{O}|U)^n \to (M|U))$).

Show that for any $\mathcal{O}$-Modules F, G and for each point $x \in X$ there is a natural morphism

$$(*) \qquad (\underline{\mathrm{Hom}}_{\mathcal{O}}(F, G))_x \to \mathrm{Hom}_{\mathcal{O}_x}(F_x, G_x),$$

which is in general neither injective nor surjective. Prove however that if F is of finite type, then the morphism (*) is always injective, while if F is of finite presentation it is bijective.

13. Let $(X, \mathcal{O})$ be a ringed space, and M, N, P be $\mathcal{O}$-Modules. Show that there is a natural bijection

$$\mathrm{Hom}_{\mathcal{O}}(M, \underline{\mathrm{Hom}}_{\mathcal{O}}(N, P)) \cong \mathrm{Hom}_{\mathcal{O}}(M \otimes_{\mathcal{O}} N, P).$$

[Compare 3. Ex. 4 and 4.10; this adjunction can be interpreted as saying that $\underline{\mathrm{Hom}}_{\mathcal{O}}$ is an internal hom-functor in $\mathcal{O}$-Mod; see [Macl] VII. 7, Schubert, Categories, 17. 8. 1.]

Use this adjunction to simplify the proof of 5. 6.

14. If M is a Module over a ringed space $(X, \mathcal{O})$ over a ring R, we can use the notation $\underline{\mathrm{End}}_{\mathcal{O}}(M)$ for the $\mathcal{O}$-Module $\underline{\mathrm{Hom}}_{\mathcal{O}}(M, M)$. Show that $\underline{\mathrm{End}}_{\mathcal{O}}(M)$ has a natural structure of sheaf of R-algebras (multiplication being given by 'composition of endomorphisms'). (Compare 5. 2.)

Show that if N is a sheaf of R-modules, then to give N a structure of $\mathcal{O}$-Module is the same as to give a morphism

$$\mathcal{O} \to \underline{\mathrm{End}}_R(N)$$

of sheaves of R-algebras (here R also denotes the constant sheaf). Note especially the case $R = \mathbf{Z}$, when N starts just as a sheaf of abelian groups.

5 · Cohomology

We now wish to measure the lack of exactness of the global section functor $\Gamma(X, -)$; we have seen that it is left exact, but need not take a sheaf epimorphism into a surjective map of sections.

We first consider the problem in the general setting of homological algebra: we wish to mend the lack of right exactness of a left exact functor between abelian categories. This leads us to define injective objects, and to show that they can be used to define the right derived functors of our functor, which fit into a long exact sequence extending the left exact sequence it produces. The right derived functors have a suitable universal property, which is used to obtain identities concerning composite functors.

We next apply this procedure to the case of sheaves. Having verified that there are enough injective sheaves, we deduce the existence of cohomology functors fitting into a long exact sequence. The general method also yields the higher direct images of a morphism, which generalise the cohomology groups, but may be expressed in terms of them. We investigate the processes of changing structure sheaves and base rings, and summarise an alternative approach to this universal cohomology theory, using flasque sheaves.

Finally we give an alternative and more computable version of cohomology, the Čech theory, which agrees with the universal theory in some useful cases. In particular we obtain a reinterpretation of the picard group of a ringed space as a cohomology group.

5.1 Injective objects

1.1 Let K be an abelian category (such as Abgp, R-mod, or $\mathcal{O}$-Mod). For $E \in \operatorname{Ob} K$ there is a contravariant functor

$$\operatorname{Hom}(-, E) : K \to \text{Abgp} : F \mapsto \operatorname{Hom}(F, E).$$

Proposition. The functor $\mathrm{Hom}(-, E)$ is always right exact; that is, if $0 \to A \to B \to C \to 0$ is exact in K, then the derived sequence

$$0 \to \mathrm{Hom}(C, E) \to \mathrm{Hom}(B, E) \to \mathrm{Hom}(A, E)$$

is exact in Abgp. $\mathrm{Hom}(-, E)$ is exact iff E has the equivalent properties:

(i) given a monomorphism $A \to B$ and a morphism $A \xrightarrow{f} E$, there is an extension $\begin{array}{ccc} A & \to & B \\ \downarrow & \swarrow & \\ E & & \end{array}$ of f to a morphism $B \to E$ making the triangle commute (but not necessarily unique);

(ii) every short exact sequence $0 \to E \to A \to B \to 0$ splits (that is, is isomorphic to

$$0 \to E \xrightarrow{\amalg_1} E \oplus B \xrightarrow{\pi_2} B \to 0).$$

Proof. Easy verification, using the universal properties of epimorphisms and cokernels, in particular. Note that a sequence

$$0 \to A \xrightarrow{g} B \to C \to 0$$

splits iff $\exists h : B \to A$ such that $h \circ g = id_A$ (see [Mit] I. 19, or Freyd, Abelian Categories 2. 68). //

1.2 **Definition.** If K is an abelian category and $E \in \mathrm{Ob}K$ is such that $\mathrm{Hom}(-, E)$ is exact, we say that E is an injective object of K.

We say that K has enough injectives iff $\forall A \in \mathrm{Ob}K$ there is an injective E and a monomorphism $A \to E$; in other words, iff every object of K can be embedded in an injective.

We shall see in §5. 2 that this is a desirable property of K, and in §5. 3 that it holds for the category of Modules over any ringed space. We show first that it always holds for the category R-mod of modules over a ring R. (The result, and indeed the proof, does not require R to be commutative; we should then talk of the category of left (or right) R-modules instead).

1.3 **Lemma.** (i) A product $\Pi_{i \in I} E_i$ is injective $\iff \forall i\, E_i$ is injective.

(ii) <u>An abelian group</u> G <u>is injective (in Abgp)</u> $\Longleftrightarrow$ G <u>is divisible (that is</u> $\forall g \in G \ \ \forall n \in \mathbf{Z}^* \ \ \exists h \in G \ \ nh = g$).

Proof. (i) $\mathrm{Hom}(F, \Pi_{i \in I} E_i) \xrightarrow{\sim} \Pi_{i \in I} \mathrm{Hom}(F, E_i)$, and the RHS is exact in F iff each factor $\mathrm{Hom}(-, E_i)$ is exact.

(ii) $\Rightarrow$: Given $g \in G$, $n \in \mathbf{Z}^*$, define: $n\mathbf{Z} \to G : n \mapsto g$. The extension $\mathbf{Z} \to G$ given by 1.1(i) sends 1 to a suitable h.

$$\begin{array}{c} n\mathbf{Z} \to G \\ \downarrow \\ \mathbf{Z} \end{array}$$

$\Leftarrow$: Given $\begin{array}{c} F \xrightarrow{\phi} G \\ f\downarrow \\ H \end{array}$ with f monomorphic, we aim to apply Zorn's Lemma to the set of extensions of ϕ to subgroups of H containing F. Certainly any chain of such extensions (under $\subseteq$) is dominated by another, so there is a maximal extension, to $F \hookrightarrow F' \hookrightarrow H$, $\phi: F \to G$, $\psi: F' \to G$ say. If $h \in H \backslash F'$, consider $\mathbf{Z}h \cap F'$. If this $= \{0\}$, we may define $\psi(h)$ arbitrarily, and obtain an extension of ψ to $F' + \mathbf{Z}h = F' \oplus \mathbf{Z}h$. If it is not $\{0\}$, then it is of the form $\mathbf{Z}nh$ for some $0 \neq n \in \mathbf{Z}$; i.e. $nh \in F'$; so $\psi(nh) = n\psi(h) \in G$, and since G is divisible we may pick $g \in G$ such that $ng = \psi(nh)$ and define $\psi(h) = g$ to extend ψ to $F' + \mathbf{Z}h$. In either case this contradicts the maximality of (F', ψ) unless $F' = H$. Hence this is the case, and by 1.1(i) G is injective. //

1.4 For reasons which will become clear, we shall be interested in the abelian group $G = \mathbf{Q}/\mathbf{Z}$; by 1.3(ii) this is injective in Abgp.

Now let R be a ring, and for an abelian group G (such as $\mathbf{Q}/\mathbf{Z}$) let $\mathrm{Hom}_{\mathbf{Z}}(R, G)$ be the R-module of abelian group morphisms $R \to G$, with the action of r given by

$$(rf)(s) = f(rs) \text{ for } r, s \in R \text{ and } f \in \mathrm{Hom}_{\mathbf{Z}}(R, G).$$

1.5 **Lemma.** (i) <u>For</u> $F \in R\text{-mod}$, $G \in \mathrm{Abgp}$ <u>there is a natural isomorphism of abelian groups</u>

$$\mathrm{Hom}_R(F, \mathrm{Hom}_{\mathbf{Z}}(R, G)) \xrightarrow{\sim} \mathrm{Hom}_{\mathbf{Z}}(F, G).$$

(ii) <u>If</u> G <u>is an injective abelian group, then</u> $\mathrm{Hom}_{\mathbf{Z}}(R, G)$ <u>is an injective</u> R-<u>module.</u>

(iii) Suppose E is an injective R-module, such that for any R-module F we have:

$$\forall f \in F \ \ f \neq 0 \Rightarrow \exists \phi \in \mathrm{Hom}_R(F, G) \text{ such that } \phi(f) \neq 0. \qquad (*15)$$

Then any R-module can be embedded in an injective (namely a product of copies of E).

Proof. (i) It is $\phi \mapsto (f \mapsto \phi(f)(1))$ for $f \in E$, with inverse $\psi \mapsto (f \mapsto (r \mapsto \psi(rf)))$ for $f \in F$, $r \in R$.

(ii) By (i), for $E = \mathrm{Hom}_{\mathbf{Z}}(R, G)$, the functor $\mathrm{Hom}_R(-, E)$ is naturally isomorphic to the functor $\mathrm{Hom}_{\mathbf{Z}}(-, G)$, which is exact since G is injective in Abgp.

(iii) The embedding we wish to use is

$$\Phi : F \to \Pi_{\phi \in \mathrm{Hom}_R(F, E)} E : f \mapsto (\phi(f))_{\phi \in \mathrm{Hom}(F, E)}$$

The RHS is injective by 1.3(i), and the R-module morphism Φ is monomorphic iff its kernel is zero i.e. iff $\forall f \in F \ \ f \neq 0 \Rightarrow f \notin \mathrm{Ker}\, \Phi$, which is the given condition. //

1.6 **Remark.** Those familiar with Commutative Algebra will recognise (i) as a special case of $\mathrm{Hom}_R(-, \mathrm{Hom}_S(-, -)) \tilde{\to} \mathrm{Hom}_S(- \otimes_R -, -)$.

1.7 **Theorem.** For a ring R, the category R-mod has enough injectives.

Proof. We shall apply 1.5(iii) to the R-module $E = \mathrm{Hom}_{\mathbf{Z}}(R, \mathbf{Q}/\mathbf{Z})$. By 1.3(ii) and 1.5(ii) E is injective. To check (*15), let $0 \neq f \in F \in \mathrm{Ob}(R\text{-mod})$. We first produce a **Z**-module morphism $\psi : F \to \mathbf{Q}/\mathbf{Z}$ such that $\psi(f) \neq 0$. We can find a non-zero **Z**-module morphism $\mathbf{Z}f \to \mathbf{Q}/\mathbf{Z}$, for if f has infinite order then $f \mapsto \frac{1}{2} + \mathbf{Z}$ will work, while if f has order $n \in \mathbf{N}^*$ then $f \mapsto \frac{1}{n} + \mathbf{Z}$ will work; by the injectivity of $\mathbf{Q}/\mathbf{Z}$ this extends to a **Z**-module morphism $\psi : F \to \mathbf{Q}/\mathbf{Z}$ such that $\psi(f) \neq 0$.

But now the formula of 1.5(i) produces a corresponding $\phi \in \mathrm{Hom}_R(F, E)$ such that $\phi(f) \neq 0$.

Hence 1.5(iii) applies and we are done. //

5.2 Derived functors

The advantage of abelian categories with enough injectives is that we can deal well with left exact functors between them.

We shall now consider some results which belong to a course on Homological Algebra. The results are obtained in an abelian category, since we wish to apply them to the category of sheaves and the section functor, but they may be easier to prove in a category like R-mod, where objects have elements; in general such a proof will suggest the proof appropriate to a more general abelian category.

2.1 **Definitions.** Let K be an abelian category. A right-co-complex $L^\cdot$ (or just complex) in K is an $\mathbf{N}$-indexed family of objects $\{L^n;\ n \in \mathbf{N}\}$ together with morphisms $d_n : L^n \to L^{n+1}$ $(n \in \mathbf{N})$ such that $\forall n \in \mathbf{N}$ $d_{n+1} \circ d_n = 0$. $L^\cdot$ is often written $L^0 \to L^1 \to L^2 \to \ldots$. The cohomology of the complex $L^\cdot$ consists of the objects

$$H^n(L^\cdot) = \operatorname{Ker} d_n / \operatorname{Im} d_{n-1} \quad (\text{for } n \in \mathbf{N};\ \text{convene } L^{-1} = 0)$$

of K; we sometimes write $H^*(L^\cdot) = \{H^n(L^\cdot);\ n \in \mathbf{N}\}$.

The complex $L^\cdot$ is exact (or acyclic) iff $\forall n \in \mathbf{N}^*$ (that is, $n \geq 1$) $H^n(L^\cdot) = 0$; that is, iff

$$0 \to H^0(L^\cdot) \to L^0 \to L^1 \to \ldots$$

is an exact sequence. If $A \in \operatorname{Ob} K$, a complex over A is a complex $L^\cdot$ with $H^0(L^\cdot) \cong A$; if this $L^\cdot$ is exact, it is called a resolution of A, and if also $\forall n \in \mathbf{N}$ L^n is injective, then $L^\cdot$ is an injective resolution of A.

A morphism of complexes $g : L^\cdot \to M^\cdot$ is given by K-morphisms $g_n : L^n \to M^n$ for each $n \in \mathbf{N}$ such that $\forall n \in \mathbf{N}$ the square

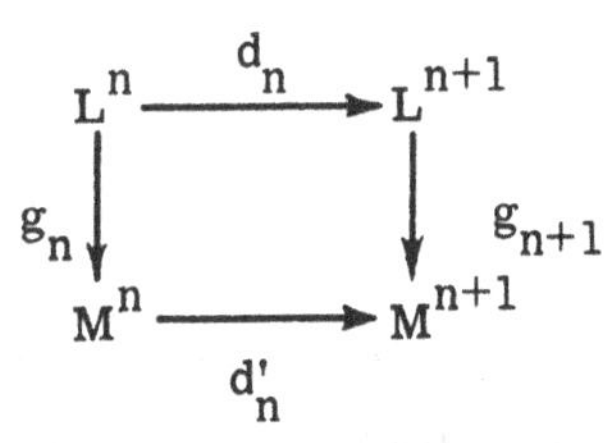

commutes (that is, $d'g = gd$). Two morphisms $g, h : L^\cdot \to M^\cdot$ are (chain) homotopic, written $g \simeq h$, iff there are K-morphisms

$k_n : L^{n+1} \to M^n$ for $n \in \mathbf{N}$ such that $\forall n \in \mathbf{N}$ $d'_{n-1}k_{n-1}+k_n d_n = g_n - h_n$ (mnemonic:

$$\begin{array}{ccccc} & & L^n & \longrightarrow & L^{n+1} \\ & \swarrow & \downdownarrows & \swarrow & \\ M^{n-1} & \longrightarrow & M^n & & \end{array}$$

).

2.2 **Proposition.** *A morphism of complexes* $g : L^\cdot \to M^\cdot$ *induces morphisms of cohomology* $H(g) = g^* : H^*(L^\cdot) \to H^*(M^\cdot)$ *in a functorial way. Homotopic maps induce the same cohomology morphism.*

Proof. This is easy to prove for K a module category, and such a proof can then be used to construct a proof in any abelian category.

In the diagram

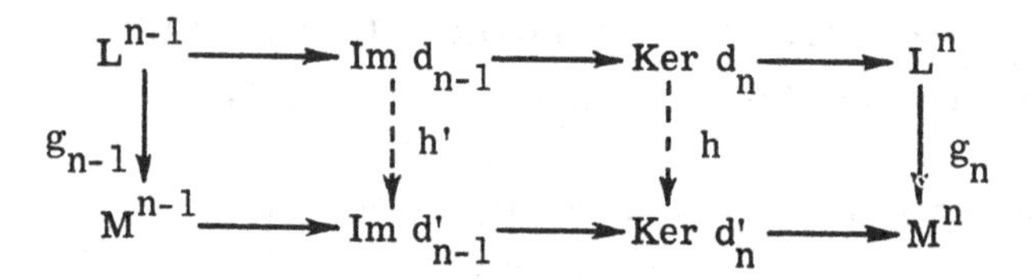

the morphisms h, h' can be constructed in turn, using the universal properties of Ker and Im; hence h induces $g^*_n : H^n(L^\cdot) \to H^n(M^\cdot)$.

If $g \simeq g' : L^\cdot \to M^\cdot$, then in the diagram

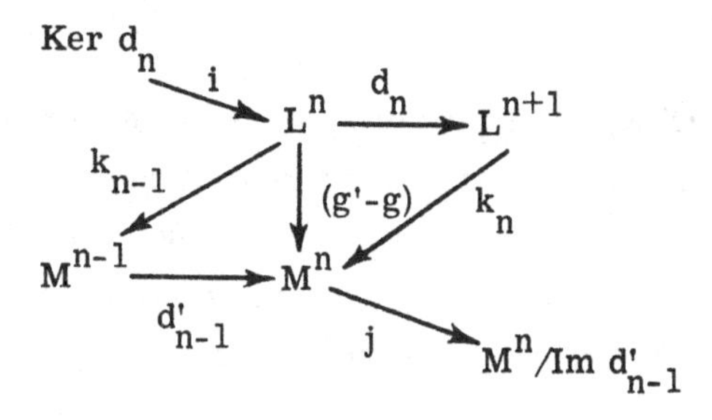

we have $j(g'-g)i = j(kd+dk)i = jk\underline{di} + \underline{jd}ki = 0$ (omitting subscripts for ease of writing), whence the result. //

2.3 **Proposition.** *If* $L^\cdot$ (resp. $M^\cdot$) *is an injective resolution of* A (resp. B), *then any morphism* $f : A \to B$ *can be lifted to a morphism* $g : L^\cdot \to M^\cdot$ (*such that* $f = g^* : H^0(L^\cdot) \to H^0(M^\cdot)$ *up to isomorphism*), *and any two such liftings are homotopic.*

Proof. We construct g_n by induction on n. If this has been done up to g_{n-1} $(n > 0)$ we have a commutative diagram

$$\begin{array}{ccccc} L^{n-1} & \longrightarrow & L^{n-1}/\mathrm{Im}\, d_{n-2} & \xrightarrow{\quad i \quad} & L^n \\ {\scriptstyle g_{n-1}}\downarrow & & \downarrow {\scriptstyle g'} & & \downarrow {\scriptstyle g_n} \\ M^{n-1} & \longrightarrow & M^{n-1}/\mathrm{Im}\, d'_{n-2} & \xrightarrow{\quad j \quad} & M^n \end{array}$$

in which i and j are injective by the exactness of $L^{\cdot}$ and $M^{\cdot}$ and g' exists since $d'_{n-1}g_{n-1}d_{n-2} = d'_{n-1}d'_{n-2}g_{n-2} = 0$. Hence since M^n is an injective object, jg' can be extended to a morphism $g_n : L^n \to M^n$ by 1.1. The outside square commutes, so g is a morphism of complexes.

To start the induction we use the same argument on the diagram

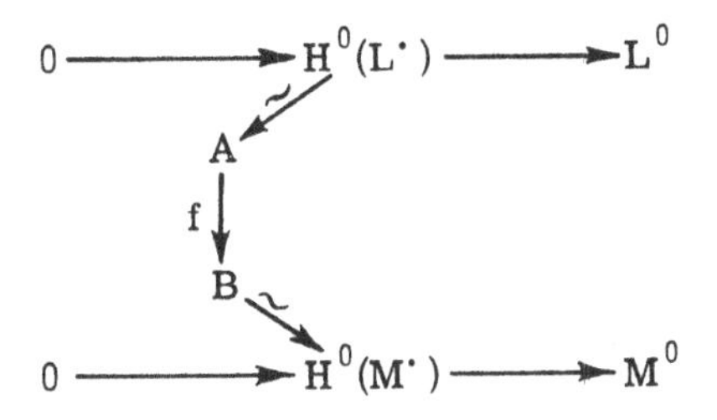

If also $h = \{h_n;\ n \in \mathbf{N}\}$ makes all the diagrams commute, we can construct a homotopy $h \simeq g$ by induction, using the diagram

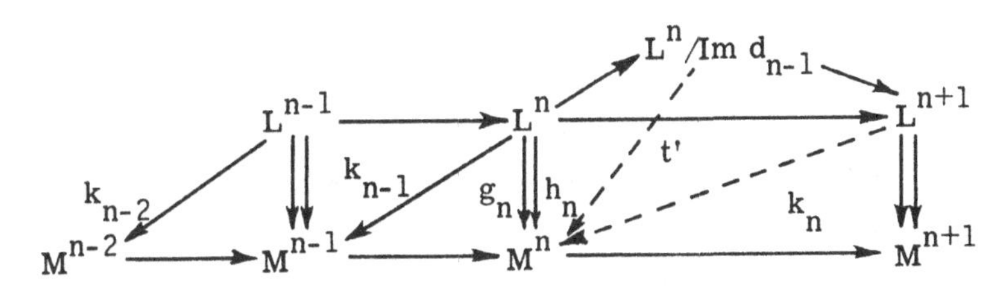

If we assume that $dk_{n-2} + k_{n-1}d = g_{n-1} - h_{n-1}$, then the map $t = g_n - h_n - d_{n-1}k_{n-1}$ is such that

$$td = gd - hd - dkd = gd - hd - d(g - h - dk) = 0;$$

hence t factors through $L^n/\mathrm{Im}\, d_{n-1}$ as t'. Since M^n is injective, t' extends to k_n such that

$$k_n d = g_n - h_n - dk_{n-1}$$

as required. //

2.4 **Remark.** A closer examination shows that we have used only the exactness of $L^\cdot$ and the injectivity of $M^\cdot$, so that a more general proposition holds.

2.5 **Corollary.** *If* K *has enough injectives (1.2), then every object* A *of* K *has an injective resolution, and any two injective resolutions* $L^\cdot$ *and* $M^\cdot$ *of* A *are homotopy equivalent: that is, there are morphisms* $L^\cdot \underset{h}{\overset{g}{\rightleftarrows}} M^\cdot$ *such that* $g \circ h \simeq id_{M^\cdot}$ *and* $h \circ g \simeq id_{L^\cdot}$.

Proof. By 1.2 A embeds in an injective object L^0, with cokernel A^0 say:

$$0 \to A \to L^0 \searrow A^0$$

Now embed A^0 in an injective L^1 with cokernel A^2, say, to get a diagram

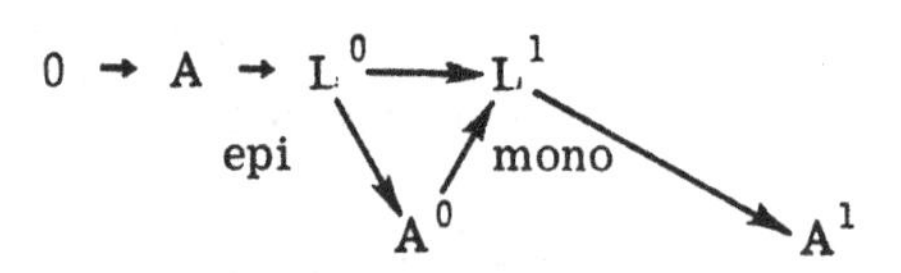

in which the top row is exact; proceeding by induction, we construct an injective resolution $L^\cdot$ of A. If $M^\cdot$ is another, then id_A lifts by 2.3 to morphisms $g : L^\cdot \to M^\cdot$ and $h : M^\cdot \to L^\cdot$, and both $h \circ g$ and $id_{L^\cdot}$ are lifts of id_A to morphisms $L^\cdot \to L^\cdot$; hence by 2.3 $h \circ g \simeq id_{L^\cdot}$. Similarly $g \circ h \simeq id_{M^\cdot}$. //

2.6 **Construction.** Let $F : K \to K'$ be a left exact functor, where K, K' are abelian categories, and K has enough injectives. Given $A \in \text{Ob}K$, let $L^\cdot$ be an injective resolution of A (2.5), and let

$$(R^nF)(A) = H^n(FK^\cdot)$$

be the cohomology of the complex $\{F(K^n), F(d_n); n \in \mathbf{N}\}$. Then the $R^nF : K \to K'$ for $n \in \mathbf{N}$ are the right derived functors of F. To define R^nF on a map $f : A \to B$ in K, lift f to a morphism of complexes $g : L^\cdot \to M^\cdot$, where $L^\cdot$, $M^\cdot$ are injective resolutions of A and B respectively (2.3), and obtain a map

$$R^nf : R^nF(A) = H^n(FL^\cdot) \xrightarrow{F(g)^*} H^n(FM^\cdot) = R^nF(B)$$

by 2.2.

2.7 **Proposition.** (a) The R^nF $(n \geq 0)$ are well defined; that is, they are independent of the choices of injective resolutions and of maps between them.

(b) R^0F is naturally isomorphic to F.

(c) If A is injective, then for $n > 0$ $R^nF(A) = 0$.

Proof. (a) Any two liftings of $f : A \to B$ to morphisms of injective resolutions $g, g' : L^\cdot \to M^\cdot$ are homotopic by 2.3; hence so are $F(g)$ and $F(g')$ (use the $F(k_n)$), which thus give the same map $R^nFA \to R^nFB$ by 2.2. The functoriality of R^nF (using fixed resolutions) follows.

Apply this to $A = B$, $f = id_A$, $L^\cdot$ and $M^\cdot$ two injective resolutions of A; we deduce that id_A lifts to unique isomorphisms $H^n(FL^\cdot) \xrightarrow{\sim} H^n(FM^\cdot)$ (they are isomorphisms by functoriality, since f is). Hence R^nFA is well-defined up to canonical isomorphism, as required.

(b) Since F is left exact, its effect on an injective resolution $0 \to H^0(L^\cdot) \to L^0 \to L^1 \to \dots$ of $A \cong H^0(L^\cdot)$ is to produce an exact sequence $0 \to FA \to FL^0 \to FL^1$. Hence

$$H^0(FL^\cdot) = \operatorname{Ker}(FL^0 \to FL^1) \cong FA.$$

(c) If A is injective, it has the injective resolution

$$\begin{array}{ccccccccc} 0 \to A & \xrightarrow{id} & A & \to & 0 & \to & 0 & \to & \dots \\ & & \| & & \| & & \| & & \\ & & L^0 & & L^1 & & L^2 & & \dots \end{array}$$

On applying F to $L^{\cdot}$ we get the complex $FA \to 0 \to 0 \to \dots$ which has

$$R^n FA = H^n(FL^{\cdot}) = \begin{cases} FA & n = 0 \\ 0 & n > 0. \end{cases} \quad //$$

2.8 **Remark.** We must be functorial, and ensure in 2.7(a) that the $R^n FA$ are well-defined up to canonical isomorphism; hence for example automorphisms of A will induce automorphisms of $R^n FA$ so as to give a group homomorphism

$$\mathrm{Aut}(A) \to \mathrm{Aut}(R^n FA).$$

2.9 **Definition.** Let K, K' be abelian categories and let $a \in \mathbf{N} \cup \{\infty\}$ (where by convention $\forall n \in \mathbf{N}\ n < \infty$). A <u>$\partial$-functor</u> $T^{\cdot} : K \to K'$ is a sequence of functors $\{T^n : K \to K';\ 0 \le n < a\}$ together with an assignment to each short exact sequence

(*S) $\quad 0 \to A \to B \to C \to 0$

in K of a collection of morphisms $\partial = \partial_T : T^{n-1}C \to T^n A$ $(0 < n < a)$ such that

(i) if
$$\begin{array}{ccccccccc} 0 & \to & A & \to & B & \to & C & \to & 0 \\ & & \downarrow f & & \downarrow & & \downarrow g & & \\ 0 & \to & A' & \to & B' & \to & C' & \to & 0 \end{array}$$
commutes in K and has exact rows, then the corresponding diagrams
$$\begin{array}{ccc} T^{n-1}C & \xrightarrow{\partial} & T^n A \\ T^{n-1}g \downarrow & & \downarrow T^n f \\ T^{n-1}C' & \xrightarrow{\partial} & T^n A' \end{array}$$
commute (in other words, ∂ is 'natural'); and

(ii) whenever (*S) is exact, the associated long sequence

(*L) $\quad 0 \to T^0 A \to T^0 B \to T^0 C \xrightarrow{\partial} T^1 A \to \dots \to T^{n-1}C \xrightarrow{\partial} T^n A \to \dots$
(for $n < a$)

is a complex (the composite of two successive morphisms is zero).

The ∂-functor is called <u>exact</u> iff for any sequence (*S) the corresponding sequence (*L) is always exact.

A <u>morphism</u> of ∂-functors (with the same a) $S^{\cdot} \to T^{\cdot}$ is given by a sequence of natural transformations $\{S^n \to T^n;\ 0 \le n < a\}$ such that for any short exact sequence (*S) the diagrams

$$\begin{array}{ccc} S^{n-1}C & \xrightarrow{\partial_S} & S^nA \\ \downarrow & & \downarrow \\ T^{n-1}C & \xrightarrow{\partial_T} & T^nA \end{array}$$

commute; hence this gives a morphism of complexes between the long sequences (*L).

If $F : K \to K'$ is a functor, a _∂-functor over_ F is a ∂-functor $\{T^n, \partial_T\}$ together with a natural isomorphism $F \xrightarrow{\sim} T^0$; hence if T is exact, F is left exact.

2.10 Theorem. _If_ $F : K \to K'$ _is a left exact functor between abelian categories, where_ K _has enough injectives, then the sequence of functors_ $R^{\cdot}F = \{R^nF;\ n \in \mathbf{N}\}$ _forms an exact ∂-functor over_ F.

Proof. We have to show that there is a natural assignment to each short exact sequence

$$0 \to A \to B \to C \to 0$$

in K of a long exact sequence

$$0 \to FA \to FB \to FC \to R^1FA \to \ldots \to R^{n-1}FC \to R^nFA \to R^nFB \to R^nFC \to \ldots .$$

We first need two Lemmas.

2.11 Lemma. _If_ $F : K \to K'$ _is a left exact functor between abelian categories and_ $0 \to A \to B \to C \to 0$ _is a split exact sequence in_ K, _then_

$$0 \to FA \to FB \to FC \to 0$$

is also (split) exact. (Hence $F(A \oplus C) \cong FA \oplus FC$.)

Proof. $0 \to A \xrightarrow{f} B \xrightarrow{g} C \to 0$ is split exact iff $A \to B \to C$ is exact and ∃ a diagram $A \xleftarrow{f'} B \xleftarrow{g'} C$ such that $f' \circ f = id_A$ and $g \circ g' = id_C$.

Then $0 \to FA \to FB \to FC \to 0$ also has these properties, and so is split exact too. //

2.12 **Lemma.** Suppose $L^\cdot \to M^\cdot \to N^\cdot$ are morphisms of complexes in an abelian category K such that $\forall n \in \mathbf{N}$

$$0 \to L^n \to M^n \to N^n \to 0$$

is exact (we call this a short exact sequence of complexes). Then there is a collection of morphisms $\partial : H^n(N^\cdot) \to H^{n+1}(L^\cdot)$ $(n \in \mathbf{N})$ such that the sequence

$$(*LH) \quad 0 \to H^0(L^\cdot) \to H^0(M^\cdot) \to H^0(N^\cdot) \xrightarrow{\partial} H^1(L^\cdot) \to \ldots \to H^n(N^\cdot) \xrightarrow{\partial} H^{n+1}(L^\cdot) \to \ldots$$

is exact. Moreover ∂ is natural in the sense that if
$$\begin{array}{ccccccccc} 0 & \to & L^\cdot & \to & M^\cdot & \to & N^\cdot & \to & 0 \\ & & \downarrow & & \downarrow & & \downarrow & & \\ 0 & \to & L^\cdot_1 & \to & M^\cdot_1 & \to & N^\cdot_1 & \to & 0 \end{array}$$
is a commutative diagram of morphisms of complexes, with exact rows, then the induced morphism between the long exact sequences (*LH) is a morphism of complexes; that is $\forall n \in \mathbf{N}$

$$\begin{array}{ccc} H^n(N^\cdot) & \xrightarrow{\partial} & H^{n+1}(L^\cdot) \\ \downarrow & & \downarrow \\ H^n(N^\cdot_1) & \xrightarrow{\partial_1} & H^{n+1}(L^\cdot_1) \end{array}$$

commutes (where ∂_1 is the morphism constructed from the lower exact row).

Proof. This result should be familiar to those who have learnt some Algebraic Topology, perhaps only in the category $K = $ Abgp or R-mod for some ring R. The proof is rather tedious; it may be reconstructed for an arbitrary K from the proof for $K = $ Abgp in Spanier, Algebraic Topology, 4.5.4, using the techniques of 2.1-2.5 and if necessary, [Macl] VIII. 4. //

Proof of 2.10. Let $L^\cdot$ (respectively $N^\cdot$) be an injective resolution of A (respectively C). We first construct a complex of injective objects $M^\cdot$ over B and a short exact sequence of complexes

$0 \to L^\cdot \to M^\cdot \to N^\cdot \to 0$. If this has been done up to the situation in the diagram

$$\begin{array}{ccccccccc} 0 & \to & L^{n-1} & \to & M^{n-1} & \to & N^{n-1} & \to & 0 \\ & & \downarrow & & \downarrow & & \downarrow & & \\ 0 & \to & L^{n} & \to & M^{n} & \to & N^{n} & \to & 0 \end{array}$$

for $n \in \mathbf{N}$ (where we make the convention $L^{-1} = A$, $M^{-1} = B$, $N^{-1} = C$) then we can construct

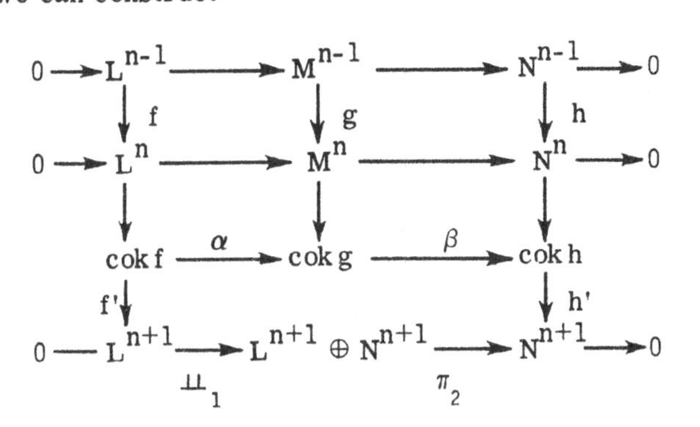

(f', h' arising from f, h by the universal property of cokernels), and it is (fairly) easy to check that the sequence of cokernels is short exact (for instance, using 'members' as in [Macl] VIII §4). We let $M^{n+1} = L^{n+1} \oplus N^{n+1}$, which is injective by 1.3(i), and define $g' : \operatorname{cok} g \to M^{n+1}$ by letting the diagram

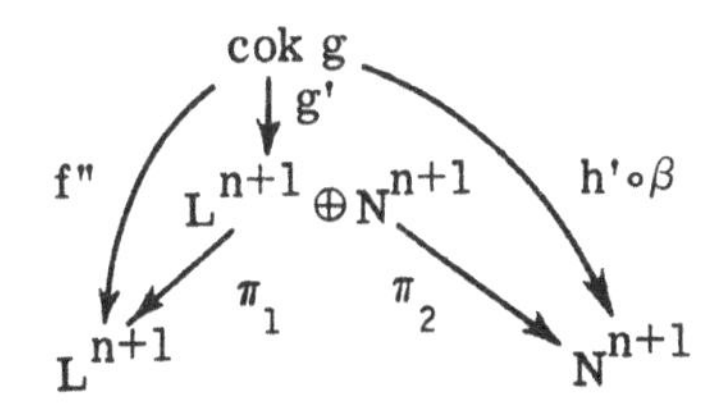

commute, where f'' is an extension of f' to $\operatorname{cok} g$, which exists by 1.1(i) since L^{n+1} is injective and α is monomorphic. It is easy to see that $M^\cdot$ is a complex and that $0 \to I^\cdot \to M^\cdot \to N^\cdot \to 0$ is a short exact sequence of complexes.

Now since $L^\cdot$ and $N^\cdot$ are exact, the long exact sequence of cohomology of 2.10 reduces to a collection of exact pieces

$$0 \to H^n(M^\cdot) \to 0 \quad \text{for} \quad n \in \mathbf{N}^*$$

which shows that $\forall n \geq 1 \;\; H^n(M^\cdot) = 0$, so that $M^\cdot$ is exact and thus is an injective resolution of B; hence we may use $M^\cdot$ to compute $R^nF(B)$.

Since each of the short exact sequences

$$0 \to L^n \to M^n \to N^n \to 0 \quad (n \in \mathbf{N})$$

is split exact (as it must be by 1.1(ii)), by Lemma 2.11 the sequence

$$0 \to FL^\cdot \to FM^\cdot \to FN^\cdot \to 0$$

of complexes is exact and so yields by Lemma 2.12 a long exact sequence

$$\ldots \to H^n(FL^\cdot) \to H^n(FM^\cdot) \to H^n(FN^\cdot) \to H^{n+1}(FL^\cdot) \to \ldots$$

which is the required sequence by 2.6 and 2.7. The naturality of ∂ follows from the corresponding property in 2.12. //

2.13 Theorem. <u>Let</u> $F : K \to K'$ <u>be a left exact functor between abelian categories, where</u> K <u>has enough injectives. Then the</u> ∂-<u>functor</u> $R^\cdot F$ <u>has the following universal property. Suppose that</u> $\{G^n, 0 \leq n < a\}$ <u>is a</u> ∂-<u>functor over</u> F (<u>where</u> $a \in \mathbf{N} \cup \{\infty\}$); <u>then there is a unique morphism of</u> ∂-<u>functors</u>

$$\{R^nF;\; 0 \leq n < a\} \to \{G^n;\; 0 \leq n < a\}$$

<u>such that the triangle</u>

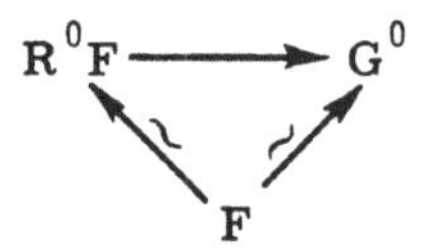

<u>commutes. If furthermore</u> $G^\cdot$ <u>is exact and</u> effaceable, <u>that is for any injective object</u> E <u>of</u> K <u>we have</u>

$$G^nE = 0 \quad \text{for} \quad 0 < n < a,$$

<u>then the morphism</u> $R^\cdot F \to G^\cdot$ <u>is an isomorphism of</u> ∂-<u>functors over</u> F.

2.13½ **Remark.** Hence the derived functor $R^{\cdot}F$ is characterised up to natural isomorphism as the exact effaceable ∂-functor over F. Conversely any exact effaceable ∂-functor $T^{\cdot}$ (sometimes called a cohomological ∂-functor) is (up to natural isomorphism) the derived functor of its T^0, and is the universal ∂-functor over T^0 (in the sense of the theorem).

Proof of 2.13. For an object A of K we construct the morphisms $R^nFA \to G^nA$ by induction on n; for $n = 0$ we compose the isomorphisms which show that $R^{\cdot}F$ and $G^{\cdot}$ are each ∂-functors over F. Embed A in an injective object E of K to obtain a short exact sequence

$$0 \to A \to E \to B \to 0$$

say. 2.7(c), the induction hypothesis and the long sequences of this exact sequence give us a diagram

$$\begin{array}{ccccccc} & 0 \longrightarrow & R^{n-1}FB & \xrightarrow{f} & R^nFA & \longrightarrow & 0 \\ & & \downarrow & & \downarrow g & & \\ G^{n-1}E \longrightarrow & & G^{n-1}B & \xrightarrow{h} & G^nA & \longrightarrow & G^nE \end{array}$$

with an exact top row; hence f is an isomorphism, and so there is a unique map g making the diagram commute. Straightforward arguments show that this is independent of the choice of E and that it defines a morphism of ∂-functors.

If $G^{\cdot}$ is exact and effaceable, the same diagram shows, by induction on n, that g is an isomorphism (since then h is). //

2.14 **Corollary.** Let $K \xrightarrow{F} K' \xrightarrow{G} K''$ be functors between abelian categories where K and K' each has enough injectives. Suppose that

(i) G is left exact

and (ii) F is exact and transforms injectives in K into G-acyclic objects; that is, whenever E is injective in K we have

$$R^nG(FE) = 0 \text{ for } n > 0.$$

Then there is a natural isomorphism of ∂-functors

$$R^{\cdot}(G \circ F) \cong (R^{\cdot} G) \circ F.$$

Proof. Apply 2.13: by the conditions, $(R^{\cdot} G) \circ F$ is an exact effaceable ∂-functor over $G \circ F$ and so is its derived functor. //

2.15 **Corollary.** Let $K \xrightarrow{F} K' \xrightarrow{G} K''$ be functors between abelian categories where K and K' each has enough injectives. Suppose that

(i) F is left exact

and (ii) G is exact.

Then there is a natural isomorphism of ∂-functors

$$R^{\cdot}(G \circ F) \cong G \circ R^{\cdot} F.$$

Proof. Apply 2.13: $G \circ R^{\cdot} F$ is an exact effaceable ∂-functor over $G \circ F$. //

2.16 **Example.** We already have enough machinery to give the definition of one cohomology theory.

Let G be a group and $\mathbf{Z}G$ its (integral) group ring. The category $(\mathbf{Z}G)$-mod is the category of G-modules (abelian groups with an action of G), and has enough injectives by 1.7. The functor

$$(-)^G : \mathbf{Z}G\text{-mod} \to \text{Abgp} : A \mapsto A^G = \{a \in A;\ \forall g \in G\ \ ga = a\}$$

is left exact, and its derived functors are the cohomology of G with coefficients in A:

$$H^n(G, A) = R^n(-)^G.(A) \qquad (n \in \mathbf{N})$$

(see Shatz, Profinite Groups, Arithmetic and Geometry; or Lang, Rapport sur la cohomologie des groupes).

More generally, let k be any (commutative) ring and kG the group algebra of G over k (k is often a field). We have a diagram of functors

$$\begin{array}{ccc} kG\text{-mod} & \xrightarrow{S'} & k\text{-mod} \\ & \searrow^{S} & \downarrow F \\ & & \mathbf{Z}\text{-mod} \end{array}$$

where $SA = A^G = \{a \in A;\ \forall g \in G \ \ ga = a\}$, $S'A$ is the same $\mathbf{Z}$-module considered with its k-module structure, and F is the forgetful functor. Then F is exact, so by 2.15

$$R^{\cdot}S \cong F \circ R^{\cdot}S'.$$

Furthermore, for $A \in \mathrm{Ob}(kG\text{-mod})$ we can recover the k-module structure on $R^nS(A)$ by considering the endomorphisms induced by the $A \to A : a \mapsto \lambda a$ for $\lambda \in k$. Hence we may compute the cohomology of G with coefficients in k:

$$H^n_k(G,\ A) = R^nS'(A)$$

by taking a resolution of A by injective ZG-modules (that is by computing $R^nS(A)$).

5.3 Sheaf cohomology

3.1 We have now constructed a general method of measuring the lack of exactness of a left exact functor; in order to apply this to the functor

$$\Gamma(X,\ -) : \mathcal{O}\text{-Mod} \to R\text{-mod}$$

for a ringed space $(X,\ \mathcal{O})$ over a ring R, we must check that $\mathcal{O}$-Mod has enough injectives.

3.2 **Lemma.** _Let_ $(X,\ \mathcal{O})$ _be a ringed space over the ring_ R. _Let_ $(M_x)_{x\in X}$ _be a family such that for each_ $x \in X$, M_x _is an_ $\mathcal{O}_x$-_module. Then there is an_ $\mathcal{O}$-_Module_ M _such that whenever_ N _is another_ $\mathcal{O}$-_Module_,

$$\mathrm{Hom}(N,\ M) \to \prod_{x\in X} \mathrm{Hom}_{\mathcal{O}_x}(N_x,\ M_x) : f \mapsto (f_x)_{x\in X}$$

is a bijection.

Proof. We can define M by its sections:

$$\Gamma(U, M) = \Pi_{x \in U} M_x \quad \text{for } U \text{ open in } X;$$

this is easily seen to define a sheaf. Alternatively, for $x \in X$ let M^x be the $\mathcal{O}$-Module with stalks

$$(M^x)_y = \begin{cases} 0 & y \neq x \\ M_x & y = x . \end{cases}$$

Then $M = \Pi_{x \in X} M^x$ is the product in $\mathcal{O}$-Mod of the M^x (4. 4. 8). //

3. 3 **Lemma.** Under the hypotheses of 3. 2, if for each $x \in X$, M_x is an injective $\mathcal{O}_x$-module, then M is an injective object in $\mathcal{O}$-Mod.

Proof. Easy from 3. 2 and 1. 1(i). //

3. 4 **Theorem.** If $(X, \mathcal{O})$ is a ringed space over a ring R, then the category $\mathcal{O}$-Mod has enough injectives.

Proof. Given an $\mathcal{O}$-Module A, for each $x \in X$ we can find an embedding (= monomorphism) $A_x \to E_x$ with E_x an injective $\mathcal{O}_x$-module (1. 7). Then 3. 2 provides an embedding $A \to E$ with E an injective $\mathcal{O}$-Module by 3. 3. //

3. 5 **Theorem.** If $(X, \mathcal{O})$ is a ringed space over a ring R, there is a universal ∂-functor (see 2. $13\frac{1}{2}$)

$$\{H^n(X, -) : \mathcal{O}\text{-Mod} \to R\text{-mod}; \ n \in \mathbf{N}\}$$

over the functor $\Gamma(X, -) : \mathcal{O}\text{-Mod} \to R\text{-mod}$. Hence an exact sequence of $\mathcal{O}$-Modules $0 \to A \to B \to C \to 0$ gives rise in a natural way to a long exact sequence:

$$0 \to \Gamma(X, A) \to \Gamma(X, B) \to \Gamma(X, C) \to H^1(X, A) \to \dots$$
$$\to H^{n-1}(X, C) \to H^n(X, A) \to H^n(X, B) \to H^n(X, C) \to \dots$$

($H^n(X, A)$ is called the cohomology of X with coefficients in A).

Proof. Set $H^n(X, -) = R^n(\Gamma(X, -))$ and apply 2.10 and 2.13. //

3.6 **Corollary.** *Suppose that* A *is an* $\mathcal{O}$*-Module with* $H^1(X, A) = 0$. *Then for any exact sequence of* $\mathcal{O}$*-Modules* $0 \to A \to B \to C \to 0$ *the induced map* $\Gamma(X, B) \to \Gamma(X, C)$ *is surjective.* //

3.7 **Theorem.** *Let* $\Phi : (X, \mathcal{O}_X) \to (Y, \mathcal{O}_Y)$ *be a morphism of ringed spaces over a ring* R. *Then there is a universal* ∂*-functor*

$$\{R^n\Phi_* : \mathcal{O}_X\text{-Mod} \to \mathcal{O}_Y\text{-Mod}\}$$

over the left exact functor Φ_*. *Hence an exact sequence*

$$0 \to A \to B \to C \to 0$$

of $\mathcal{O}_X$*-Modules gives rise in a natural way to an exact sequence of* $\mathcal{O}_Y$*-Modules:*

$$0 \to \Phi_*A \to \Phi_*B \to \Phi_*C \to R^1\Phi_*A \to R^1\Phi_*B \to R^1\Phi_*C \to R^2\Phi_*A \to \dots$$

Proof. Φ_* is left exact by 4.4.13 and 3.7.6. By 3.4 we may apply 2.10 and 2.13. //

3.8 **Remark.** Let P be a topological space with just one point. Then a sheaf over P is given by just one set of sections (since for a sheaf F of abelian groups $\Gamma(F, \emptyset) = \{0\}$) which is also the stalk at $p \in P$. Hence we can make P into a ringed space over R, the structure sheaf $\mathcal{O}_P$ having stalk R. Then there is an equivalence of categories

$$\mathcal{O}_P\text{-Mod} \xrightarrow{\sim} R\text{-mod}$$

and as in 4.7.5A, if $(X, \mathcal{O})$ is a ringed space over R and $\Phi : X \to P$ is the (only) morphism, the diagram

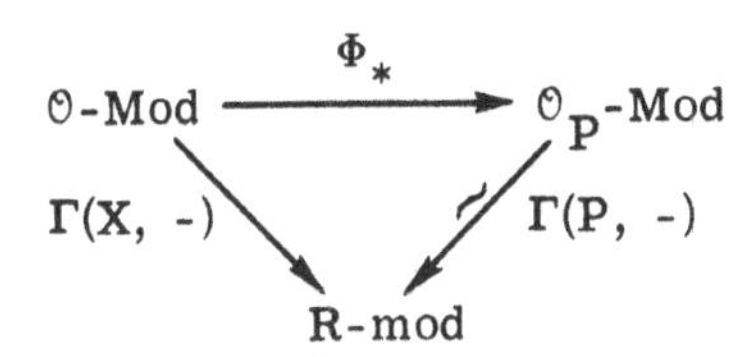

commutes. Thus the functors $R^n\Phi_*$ and $H^n(X, -)$ are the derived functors of essentially the same functor, and by the uniqueness (2.13 and 2.14) we have a natural isomorphism for $A \in \mathcal{O}$-Mod

$$\Gamma(P, R^n\Phi_*A) \cong H^n(X, A).$$

In fact, as the next result shows, we can describe $R^n\Phi_*$ for general Φ in terms of cohomology too.

3.9 **Theorem.** _If $\Phi : (X, \mathcal{O}_X) \to (Y, \mathcal{O}_Y)$ is a morphism of ringed spaces over a ring R, with underlying continuous map $\phi : X \to Y$, and A is an $\mathcal{O}_X$-Module, then $R^n\Phi_*A$ is (up to isomorphism) the sheaf associated (by sheafification) to the presheaf_

$$V \mapsto H^n(\phi^{-1}(V), A) \text{ for } V \text{ open in } Y.$$

(V is considered as a ringed space with structure sheaf $\mathcal{O}_Y|V$, so $H^n(\phi^{-1}(V), A)$ is a $\Gamma(V, \mathcal{O}_Y)$-module).

Proof. For $n \in \mathbf{N}$ let F^nA be the sheafification of the presheaf described. Then $F^n : \mathcal{O}_X\text{-Mod} \to \mathcal{O}_Y\text{-Mod}$ is easily seen to be a functor. We have $F^0 = \Phi_*$, and for an exact sequence $0 \to A \to B \to C \to 0$ in $\mathcal{O}_X$-Mod, the exact sequences

$$\ldots \to H^{n-1}(\phi^{-1}(V), C) \to H^n(\phi^{-1}(V), A) \to H^n(\phi^{-1}(V), B) \to H^n(\phi^{-1}(V), C) \to \ldots$$

of 3.5 give an exact sequence of presheaves and so by 3.6.9(ii) an exact sequence of sheaves

$$\ldots \to F^{n-1}C \to F^nA \to F^nB \to F^nC \to \ldots$$

Note that we are abusing language by writing $H^n(\phi^{-1}(V), A)$ instead of $H^n(\phi^{-1}(V), \psi^*A)$ (where $\psi : \phi^{-1}(V) \hookrightarrow X$); since ψ^* is an exact functor (3.7.16) the application of 3.5 is valid.

Hence $\{F^n;\ n \in \mathbf{N}\}$ forms an exact ∂-functor over Φ_*; also by Lemma 3.10 below, if E is injective in $\mathcal{O}_X$-Mod and U is open in X, then $E|U$ is injective in $(\mathcal{O}_X|U)$-Mod, so for $n \in \mathbf{N}^*$ $H^n(U, E|U) = 0$, so that F^nE is the zero sheaf. Hence $F^\cdot$ is also effaceable, and 2.13

shows that it is isomorphic to the derived functor $R^{\cdot}\Phi_*$ of Φ_*. //

3.10 **Lemma.** _If_ $(X, \mathcal{O})$ _is a ringed space and_ U _is open in_ X, _then for_ E _injective in_ $\mathcal{O}$-Mod, $E|U$ _is also injective in_ $(\mathcal{O}|U)$-Mod.

Proof. We give two proofs, since each illustrates a useful technique.

(a) Let A be an $(\mathcal{O}|U)$-Module. Then by 3.8.12 if A^X denotes the extension by zero of A to X, there is a bijection

$$(*10) \quad \mathrm{Hom}_{\mathcal{O}|U}(A, E|U) \xleftarrow{\sim} \mathrm{Hom}_{\mathcal{O}}(A^X, E)$$

which is natural in A. (For the map is given by restriction to U, and has inverse given by

$$\mathrm{Hom}_{\mathcal{O}|U}(A, E|U) \xrightarrow{(-)^X} \mathrm{Hom}_{\mathcal{O}}(A^X, (E|U)^X) \to \mathrm{Hom}_{\mathcal{O}}(A^X, E)$$

where the last map is derived from the morphism $(E|U)^X \to E$ of 3.8.12; all the maps are easily seen to respect the Module structures). But now the right-hand side of (*10) is exact in A by 3.8.9 and the hypothesis on E; hence the left-hand side is exact in A and so by 1.2 $(E|U)$ is injective.

(b) If E is an injective of the kind constructed in 3.3, clearly $E|U$ is also of this type and so is injective by 3.3. For general E, 3.4 shows that we can embed E in an injective E' of the type of 3.3; but then the inclusion $E \hookrightarrow E'$ splits by 1.1(ii), so $E' \cong E \oplus E''$ for some E''. Hence

$$(E'|U) \cong (E|U) \oplus (E''|U)$$

and $(E'|U)$ is injective as already remarked; hence $E|U$ is injective by 1.3(i). //

3.11 **Proposition.** _Let_ $(X, \mathcal{O})$ _be a ringed space over_ R, _and let_ $\theta : S \to R$ _be a ring morphism. Let_ Γ' _denote the composite functor_

$$\mathcal{O}\text{-Mod} \xrightarrow{\Gamma} R\text{-mod} \xrightarrow{G} S\text{-mod}$$

where $\Gamma = \Gamma(X, -)$ and G is the change-of-structure functor associated with θ. Then there is a natural isomorphism of ∂-functors:

$$R^n\Gamma' \cong G \circ R^n\Gamma = G(H^n(X, -))$$

In other words, we may compute the cohomology of an $\mathcal{O}$-Module as if its section sets were only S-modules.

Proof. G is exact; apply 2.15. //

3.12 **Remark.** The R-module structure on $H^n(X, -)$ may be recovered as follows. For an $\mathcal{O}$-Module A and for $r \in R$, the operation 'multiply by r' induces an endomorphism of A and hence an endomorphism of $H^n(X, A)$ (as S-module) which shows how r operates on $H^n(X, A)$.

Applied to the special case $S = \mathbf{Z}$, 3.11 shows that we may always regard modules of sections as just abelian groups for the purposes of computing cohomology.

3.13 **Proposition.** Let $(X, \mathcal{O})$ be a ringed space over a ring R, which satisfies the hypothesis:

(FL) for any open U in X, $\Gamma(U, \mathcal{O})$ is a torsion free abelian group.

Let Γ' be the composite functor

$$\mathcal{O}\text{-Mod} \xrightarrow{F} \mathbf{Z}_X\text{-Mod} \xrightarrow{\Gamma} \mathbf{Z}\text{-mod}$$

where $\Gamma = \Gamma(X, -)$, $\mathbf{Z}_X$ is the constant sheaf $\mathbf{Z}$ on X and F is the forgetful functor (cf. 4.4.5A). Then there is a natural isomorphism of ∂-functors

$$R^n\Gamma' \cong (R^n\Gamma) \circ F = H^n(X, F(-)).$$

In other words, we may compute the cohomology of an $\mathcal{O}$-Module as if it were just a sheaf of abelian groups; in particular we may use resolutions of injective sheaves of abelian groups.

Proof. We wish to apply 2.14; since F is exact we have the required result provided that F transforms injectives in $\mathcal{O}$-Mod into Γ-acyclic objects. We shall show that under the hypothesis (FL), F transforms injectives into injectives.

In fact, F is the 'direct image part' of a morphism of ringed spaces (over $\mathbf{Z}$):

$$\Phi : (X, \mathcal{O}) \to (X, \mathbf{Z}_X)$$

with underlying continuous map id_X and the unique possible morphism $\mathbf{Z}_X \to (id_X)_*\mathcal{O}$. This induces an adjoint pair (4.13 and 4.14):

$$\mathcal{O}\text{-Mod} \underset{\Phi^*}{\overset{\Phi_* = F}{\rightleftarrows}} \mathbf{Z}_X\text{-Mod}.$$

Now let E be injective in $\mathcal{O}$-Mod, and A a $\mathbf{Z}_X$-Module. Then by 4.14 there is a natural bijection

$$\text{Hom}(A, FE) \xrightarrow{\sim} \text{Hom}(\Phi^*A, E)$$

and by 4.13 the functor Φ^* is given by

$$\Phi^*A = A \otimes_{\mathbf{Z}_X} \mathcal{O}.$$

Under the assumption (FL), each $\Gamma(U, \mathcal{O})$ is a flat $\mathbf{Z}$-module (see for example [K], 2.2.3.1), and so (using 3.6.9(ii)) Φ^* is an exact functor. Hence Hom(A, FE) is exact in A and so FE is injective by 1.2. //

3.14 **Remark.** As in 3.12, the R-module structure on the $H^n(X, -)$ can be recovered.

3.15 **Remark.** In fact the conclusion of 3.13 holds even without the hypothesis (FL) on $(X, \mathcal{O})$. This is most easily seen by using flasque sheaves, and since the details of the proof have other applications, we give a sketch for the convenience of the reader.

A sheaf F of abelian groups on a topological space X (or more generally an $\mathcal{O}$-Module F on a ringed space $(X, \mathcal{O})$) is <u>flasque</u> (= soggy) iff for every open set U of X the restriction map $\Gamma(X, F) \to \Gamma(U, F)$

is surjective (and hence so is every restriction map of F). Then:

(a) If $0 \to F \to G \to H \to 0$ is an exact sequence of sheaves and F, G are flasque, then so is H, and the sequence

$$0 \to \Gamma(X, F) \to \Gamma(X, G) \to \Gamma(X, H) \to 0$$

is exact. (Proof by direct manipulation.)

(b) A direct summand of a flasque sheaf is flasque. (Direct proof.)

(c) Every sheaf F can be embedded in a flasque sheaf F^0; indeed F^0 can be made a functor of F as follows. Let $E \xrightarrow{p} X$ be the sheaf space of F, and let E^0 be the set E with the coarsest topology such that p remains continuous (that is, with open sets all the $p^{-1}(U)$ for U open in X). Let F^0 be the sheaf of sections of $E^0 \xrightarrow{p} X$ (2.2.C); thus for U open in X, $\Gamma(U, F^0)$ is the set $\Pi_{x \in U} F_x$ of not-necessarily continuous sections of $p : E \to X$. (Compare 3.2.) The natural map $F \to F^0$ is the required embedding.

(d) Every injective sheaf is flasque. (First proof: embed E in a flasque E^0 by (c); then E is a direct factor of E^0 by 1.1(ii) and so is flasque by (b). Second proof: for U open in X apply the exact functor $\mathrm{Hom}_{\mathcal{O}}(-, E)$ to the exact sequence

$$0 \to (\mathcal{O}|U)^X \to \mathcal{O} \to (\mathcal{O}|X\backslash U)^X \to 0$$

of 4.8.11.)

(e) After (c), by the technique of 2.5 every sheaf F has a flasque resolution

$$0 \to F \to F^0 \to F^1 \to \ldots$$

Apply $\Gamma(X, -)$ to this complex and take cohomology. Then it can be shown that the resulting functors

$$H_f^*(X, F) = H^*(\Gamma(X, F^\cdot))$$

form an exact ∂-functor with

$$\begin{aligned} H_f^0(X, F) &= \ker(\Gamma(X, F^0) \to \Gamma(X, F^1)) \\ &= \Gamma(X, F). \end{aligned}$$

Also $H_f^*(X, -)$ is effaceable, by (d), since any flasque F has the resolution $0 \to F \xrightarrow{id} F \to 0 \to 0 \to \dots$ Hence by 2.13

$$H_f^*(X, F) \cong H^*(X, F).$$

(This is the definition of sheaf cohomology used by Godement in [G].)

(f) We deduce that for any flasque sheaf F,

$$H^n(X, F) = \begin{cases} \Gamma(X, F) & n = 0 \\ 0 & n > 0; \end{cases}$$

that is, F is acyclic. In fact this follows directly from (a), (b) and (c); see [T] 3.3.1.

But now, in the situation of 3.13 (without (FL)), if M is an injective $\mathcal{O}$-Module, then it is flasque by (d), and hence F(M) is a flasque $\mathbf{Z}_X$-Module, and so acyclic by (f). But this is enough to prove 3.13 by applying 2.14.

3.16 **Exercise.** Formulate and prove the generalisation of 3.13 to the 'change of sheaf of rings from $\mathcal{O}$ to $\mathcal{O}'$' situation (giving 3.13 when $\mathcal{O}' = \mathbf{Z}_X$). Under what situations does it generalise fully to morphisms of ringed spaces, and with $R^n\Phi_*$ in place of $R^n\Gamma = H^n(X, -)$? (Hint: the spectral sequence of a composite functor: [T] 2.4.1.)

3.17 For some applications we may be interested in left exact subfunctors of Γ and their derived functors; in particular, we can obtain cohomology with supports as follows. We formulate the results for sheaves of abelian groups; the extension to ringed spaces is clear.

A set Φ of closed subsets of a topological space X is called a <u>system of supports</u> iff it satisfies the conditions:

(a) $A, B \in \Phi \Rightarrow A \cup B \in \Phi$

(b) $A \subseteq B \in \Phi$ and A closed $\Rightarrow A \in \Phi$.

For example $\Phi = \{\{x\}, \emptyset\}$ where x is a closed point; or if X is hausdorff $\Phi = \{A;\ A$ is compact$\}$.

For F a sheaf on X, the group of sections of F with Φ-support is

$$\Gamma_\Phi(F) = \{s \in \Gamma(X, F);\ |s| \in \Phi\}$$

where $|s| = \{x \in X;\ s_x \neq 0\}$. Then Γ_Φ is a left exact functor: Shv/X → Abgp and its derived functors are the cohomology of X with Φ-support

$$H^n_\Phi(X, F) = R^n\Gamma_\Phi(F).$$

For instance we obtain in this way the cohomology of X with compact support. For more details, see Swan, Theory of sheaves, [G] or Bredon, Sheaf Theory.

5.4 Čech cohomology

The functor 'Γ' : Presh/X → Abgp : F ↦ F(X) is exact (3.6.9(iii)) and so has zero right derived functor. However there is a ∂-functor Presh/X → Abgp which is of interest as it may sometimes aid in the computation of sheaf cohomology, whose existence is guaranteed by 3.5, but in a way that makes evaluation difficult.

4.1 Let I be a set. For $n \in \mathbf{N}$, let $[0, n] = \{n \in \mathbf{N};\ 0 \le m \le n\}$. An n-simplex on I is a function $\sigma : [0, n] \to I$, and the set of n-simplices is denoted by I_n.

For $0 \le m \le n+1$ there are maps ('omit the mth vertex')

$$\partial_m : I_{n+1} \to I_n : \sigma \mapsto \left(\sigma' : k \mapsto \begin{cases} \sigma(k) & \text{if } k < m \\ \sigma(k+1) & \text{if } k \ge m \end{cases}\right).$$

4.2 Let X be a topological space and let F be a presheaf of abelian groups on X (that is an object of the category Presh/X). Let $\mathfrak{U} = (U_i)_{i \in I}$ be an open cover of X. For an n-simplex $\sigma \in I_n$ let

$$U_\sigma = \cap\{U_{\sigma(m)};\ m \in [0, n]\},$$

which is an open set of X. For $n \in \mathbf{N}$ let

$$C^n(\mathfrak{U}, F) = \Pi_{\sigma \in I_n} F(U_\sigma).$$

This is an abelian group, and we use notations like (s_σ) for the element whose σ^{th} coordinate is s_σ.

The maps ∂_m of 4.1 induce

$$\partial_m : C^n(\mathfrak{U}, F) \to C^{n+1}(\mathfrak{U}, F)$$
$$(s_\sigma) \mapsto (t_{\sigma'}) \quad \text{where } t_{\sigma'} = s_{\partial_m \sigma'}$$

and we set for each n

$$d_n = \sum_{m=0}^{n+1}(-1)^m \partial_m : C^n(\mathfrak{U}, F) \to C^{n+1}(\mathfrak{U}, F).$$

It is easy to check that in this way $C^{\cdot}(\mathfrak{U}, F)$ becomes a complex, called the Čech complex belonging to $\mathfrak{U}$ and F, and its cohomology (2.1)

$$H^n(\mathfrak{U}, F) = H^n(C^{\cdot}(\mathfrak{U}, F)) \quad (n \in \mathbf{N})$$

is the Čech cohomology of F with respect to the covering $\mathfrak{U}$. The constructions of $C^{\cdot}(\mathfrak{U}, F)$ and $H^*(\mathfrak{U}, F)$ are functorial in the presheaf F.

4.3 For example, $H^0(\mathfrak{U}, F)$ is computed as the kernel of

$$\begin{array}{ccc} C^0(\mathfrak{U}, F) & \to & C^1(\mathfrak{U}, F) \\ \| & & \| \\ \Pi_{i\in I} F(U_i) & \to & \Pi_{(i,j)\in I\times I} F(U_i \cap U_j) \end{array}$$

where the lower map sends (s_i) to

$$(\rho^{U_j}_{U_i \cap U_j}(s_j) - \rho^{U_i}_{U_i \cap U_j}(s_i))$$

(compare 2.1.8). Hence if F is in fact a sheaf, we have

$$H^0(\mathfrak{U}, F) \cong F(X)$$

for any open covering $\mathfrak{U}$.

4.4 Given two open coverings $\mathfrak{U} = (U_i)_{i\in I}$, $\mathfrak{V} = (V_j)_{j\in J}$, we say that $\mathfrak{V}$ is a refinement of $\mathfrak{U}$ iff there is a refinement map $r : J \to I$ with the property that

$$\forall j \in J \quad V_j \subseteq U_{r(j)}.$$

Such a refinement map r induces a morphism of complexes

$$r' : C^{\cdot}(\mathfrak{U}, F) \to C^{\cdot}(\mathfrak{V}, F)$$

derived from the maps

$$\Pi_{\sigma \in I_n} F(U_\sigma) \to \Pi_{\tau \in J_n} F(V_\tau) : (s_\sigma) \mapsto (t_\tau)$$

where $t_\tau = \rho^{U_{r(\tau)}}_{V_\tau}(s_{r(\tau)})$, and where r also denotes the map induced by composition from J_n to I_n.

Hence a refinement map induces (2. 2) a morphism of Čech cohomology

$$\bar{r} : H^n(\mathfrak{U}, F) \to H^n(\mathfrak{V}, F) \qquad (n \in \mathbf{N}).$$

4. 5 **Lemma.** If $\mathfrak{V}$ is a refinement of $\mathfrak{U}$, and $r_1, r_2 : J \to I$ are two refinement maps, then they induce the same morphism of Čech cohomology

$$\bar{r}_1 = \bar{r}_2 : H^n(\mathfrak{U}, F) \to H^n(\mathfrak{V}, F).$$

Proof. In fact, in the notation of 4. 4, r_1' and r_2' are homotopic morphisms of complexes (2. 1), by the homotopy:

$$k_n : C^{n+1}(\mathfrak{U}, F) \to C^n(\mathfrak{V}, F)$$
$$(s_\sigma) \mapsto (t_\tau)$$

where $t_\tau = \sum_{k=0}^{n} (-1)^k \rho^{U_{\tau_k}}_{V_\tau}(s_{\tau_k})$,

where $\tau_k(m) = \begin{cases} r_1(\tau(m)) & \text{if } m \leq k \\ r_2(\tau(m-1)) & \text{if } m > k \end{cases}$ for $m \in [0, n+1]$. The result follows by 2. 2. //

4. 6 Hence the abelian groups $H^n(\mathfrak{U}, F)$ form a directed system as $\mathfrak{U}$ varies over the open covers of X, and we define the Čech cohomology of the presheaf F on X to be

$$\check{H}^n(X, F) = \varinjlim_{\mathfrak{U}} H^n(\mathfrak{U}, F)$$

(over finer and finer covers). There is a set-theoretic difficulty in that the class of open covers of X is not a set, but we may avoid this either by allowing only covers indexed by subsets of a suitably large set such as the power set of X, or by proceeding as in 4.7 below.

4.7 If $\mathcal{V}$ is a refinement of $\mathcal{U}$, the morphism

$$r' : C^{\cdot}(\mathcal{U}, F) \to C^{\cdot}(\mathcal{V}, F)$$

depends on the choice of the refinement r, although $\bar{r}$ does not. Hence we have difficulty in forming a direct system of the $C^{\cdot}(\mathcal{U}, F)$ so as to obtain an exact cohomology sequence. This can be solved by using a neat trick due to Godement ([G] §5.8).

Let $\mathcal{R}(X)$ be the set of open covers $(U_x)_{x \in X}$ of X which are indexed by X in such a way that

$$\forall x \in X \quad x \in U_x.$$

On $\mathcal{R}(X)$ we define a preorder by

$$\mathcal{V} \geq \mathcal{U} \quad \text{iff} \quad \forall x \in X \quad V_x \subseteq U_x$$

and so obtain a canonical refinement map $X \xrightarrow{id} X$ when $\mathcal{V} \geq \mathcal{U}$.

As in 4.3, $\mathcal{V} \geq \mathcal{U}$ implies that we have a morphism

$$C^{\cdot}(\mathcal{U}, F) \to C^{\cdot}(\mathcal{V}, F)$$

of complexes, and we define

$$C^{\cdot}(X, F) = \varinjlim_{\mathcal{U} \in \mathcal{R}(X)} C^{\cdot}(\mathcal{U}, F).$$

Since $\varinjlim$ is exact (see Bourbaki, <u>Algèbre</u>, Chapter II, 6.6.8 and 1.Ex.8) we find

$$\check{H}^n(X, F) = H^n(C^{\cdot}(X, F)).$$

4.8 Theorem. $\{\check{H}^n(X, -); n \in \mathbf{N}\}$ <u>is an exact</u> ∂-<u>functor</u>: Presh/X $\to$ Abgp.

Proof. An exact sequence $0 \to P \to Q \to R \to 0$ in Presh/X gives for each $\mathcal{U} \in \mathcal{R}(X)$ an exact sequence

$$0 \to C^{\cdot}(\mathcal{U}, P) \to C^{\cdot}(\mathcal{U}, Q) \to C^{\cdot}(\mathcal{U}, R) \to 0$$

as is easily checked. Applying the exact functor $\varinjlim$ we get an exact sequence of complexes

$$0 \to C^{\cdot}(X, P) \to C^{\cdot}(X, Q) \to C^{\cdot}(X, R) \to 0$$

which gives by 2.12 the required long exact sequence. //

4.9 **Exercise.** Deduce from 4.8 that $\check{H}^*(X, -)$ is the right derived functor of $\check{H}^0(X, -)$. [Hint: Use 2.13; to check that Presh/X has enough injectives, mimic 3.2 and 3.3 (compare Artin, Grothendieck Topologies, I, 2.7 and 3.2).]

4.10 **Exercise.** Show that $F \mapsto F^+$ where for U open in X

$$F^+(U) = \check{H}^0(U, F|U)$$

is a functor Presh/X $\to$ Presh/X with the properties:

(i) for any presheaf F, F^+ is a monopresheaf;

(ii) if F is a monopresheaf, F^+ is a sheaf;

(iii) for any presheaf F there is a natural morphism $F \to F^+$ such that if G is a sheaf then any morphism $F \to G$ factors through $F \to F^+$.

Deduce that F^{++} is (isomorphic to) the sheafification of F.

4.11 If we consider the restriction of $\check{H}^n(X, -)$ to the category $\mathrm{Shv}/X = \mathbf{Z}_X\text{-Mod}$ of sheaves of abelian groups on X, we find that it is not necessarily a ∂-functor (exact sequences in Shv/X are not necessarily exact in Presh/X). By 4.3 we know that if P is a sheaf on X, then

$$\check{H}^0(X, P) = \Gamma(X, P) = H^0(X, P).$$

An exact sequence

$$0 \to P \xrightarrow{f} Q \to R \to 0$$

in Shv/X has $R \cong SCok(f)$ and so gives a diagram

$$\begin{array}{ccccccc} & & & & R & & \\ & & & \nearrow & \uparrow & & \\ 0 \to P \to Q & \to & R' = PCok(f) & \to 0 & & & \end{array}$$

where the horizontal line is exact in Presh/X and the vertical map is the natural map of sheafification. Thus by 4.8 we obtain diagrams

$$\begin{array}{ccccccccc} & \Gamma(X, P) & \to & \Gamma(X, Q) & \to & \Gamma(X, R) = \check{H}^0(X, R) & & & \\ & \| & & \| & & \nearrow & & & \\ 0 \to & \check{H}^0(X, P) & \to & \check{H}^0(X, Q) & \to & \check{H}^0(X, R') & \to & \check{H}^1(X, P) & \to \ldots \end{array}$$

$$\begin{array}{ccccccc} & & & \check{H}^n(X, R) & & & \\ & & \nearrow & & & & \\ \ldots \to \check{H}^n(X, Q) & \to & \check{H}^n(X, R') & \to & \check{H}^{n+1}(X, P) & \to \ldots & \end{array}$$

with exact bottom rows.

4.12 **Lemma.** _If E is an injective sheaf of abelian groups on X then for_ $n > 0$

$$\check{H}^n(X, E) = 0$$

(_that is_, $\check{H}^*(X, -)$ _is effaceable on_ Shv/X).

Proof. We show in fact that if E is any sheaf of the kind constructed in 3.2, then for each open covering $\mathfrak{U}$ of X, the Čech complex $C^{\cdot}(\mathfrak{U}, E)$ is contractible; that is, it admits a homotopy (defined below) between the identity and zero endomorphisms. Hence $id = 0 : H^n(\mathfrak{U}, E) \to H^n(\mathfrak{U}, E)$ for $n > 0$, and so $\check{H}^n(X, E) = 0$ for $n > 0$. By 3.3 any injective sheaf E' embeds in a sheaf E of this form, and so $E = E' \oplus E''$ for some E" since E' is injective (1.1). Hence for $n > 0$

$$0 = \check{H}^n(X, E) = \check{H}^n(X, E') \oplus \check{H}^n(X, E'')$$

(this follows easily from the exact cohomology sequence of the split exact sequence $0 \to E' \to E \to E'' \to 0$) and the result follows.

Given E of the type described in 3.2 and an open covering $\mathfrak{U} = (U_i)_{i \in I}$ of X, pick $j \in I$. A suitable homotopy is then given by the morphisms

$$k_n : C^{n+1}(\mathfrak{U}, E) \to C^n(\mathfrak{U}, E) : (s_\sigma) \mapsto (s'_{\tau'})_{\tau \in I_n}$$

where $\tau'(m) = \begin{cases} j & \text{for } m = 0 \\ \tau(m-1) & \text{for } m > 0 \end{cases}$ for $m \in [0, n+1]$,
and for $\tau \in I_n$, $s'_{\tau'}$ is the image of $s_{\tau'}$ under the natural map

$$\Pi_{x \in U_{\tau'}} E_x \to \Pi_{x \in U_\tau} E_x : (e_x) \mapsto (\{\begin{matrix} e_x & \text{if } x \in U_{\tau'} \\ 0 & \text{if } x \notin U_{\tau'} \end{matrix}\})$$

(we have $U_{\tau'} \subseteq U_\tau$, and we may have to enlarge the 'domain' of the section $s_{\tau'}$). //

4.13 We see from 2.13, 4.11 and 4.12 that for a fixed space X and $a \in \mathbf{N} \cup \{\infty\}$, we can assert that for any sheaf F of abelian groups on X we have

(*413) $H^n(X, F) \cong \check{H}^n(X, F)$ for $0 \leq n < a$

provided that $\{\check{H}^n(X, -);\ 0 \leq n < a\}$ forms an exact ∂-functor, which will be true if the following condition holds:

(a) if R' is any presheaf on X with sheafification R, then the induced Čech cohomology map

$$\check{H}^n(X, R') \to \check{H}^n(X, R)$$

is an isomorphism for $0 \leq n < (a - 1)$ (where $\infty - 1 = \infty$).

If we let S be a presheaf such that the sequence $0 \to R' \to R \to S \to 0$ is exact in Presh/X, we see that S has as sheafification the zero sheaf, and there is an exact sequence

$$\ldots \to \check{H}^{n-1}(X, S) \to \check{H}^n(X, R') \to \check{H}^n(X, R) \to \check{H}^n(X, S) \to \ldots$$

It follows that the condition (a) is equivalent to

(b) if S is a presheaf on X with zero sheafification, then for $0 < n < (a - 1)$

$$\check{H}^n(X, S) = 0.$$

This condition holds, and hence so does (*413), in a number of interesting cases:

(i) for any X for $a = 2$ (see 4.14 below);

(ii) for $a = \infty$ providing X is paracompact (see [G], 5.10.1); [or more generally, for $a = \infty$ and any X providing we replace H^n, $\check{H}^n$ by their versions 'with supports in a paracompactifying family' (cf. 3.16; see also Swan, Theory of Sheaves, VIII)];

(iii) for $a = \infty$ if X is a scheme and provided we restrict our attention to quasi-coherent $\mathcal{O}_X$-Modules (see [EGA III] 1.4.1).

More generally, we have in any case a spectral sequence

$$H^p(\mathfrak{U}, \mathcal{H}^q(F)) \Rightarrow H^{p+q}(X, F)$$

for any open cover $\mathfrak{U}$ of X, where $\mathcal{H}^q(F)$ denotes the presheaf $U \mapsto H^q(U, F)$; and (i)-(iii) above can be regarded as cases where it degenerates. See [T], 3.8.1; [G] 5.9.1; Artin, Grothendieck Topologies, II §3.

For the connections between singular cohomology, Alexander-Spanier cohomology and sheaf cohomology (Čech or Grothendieck (derived functor) version), the reader is referred to Spanier, Algebraic Topology; particularly 6.8.8, 6.9.1, 6.9.5, 6.9.7. In general terms these results state that for suitably nice topological spaces X (such as topological manifolds), the cohomology of a constant sheaf G on X is isomorphic to the singular cohomology of X with coefficients in G.

More generally, if G now denotes a locally constant sheaf, then the sheaf cohomology of G gives the cohomology of X with the local system of coefficients G (see Spanier, Chapter 6, Exercise F and Chapter 5, Exercise J).

4.14 **Theorem.** _If X is any topological space and S is a presheaf of abelian groups on X with zero sheafification, then_

$$\check{H}^0(X, S) = 0.$$

Hence as in 4.13, $(\check{H}^0, \check{H}^1)$ form an exact effaceable ∂-functor on Shv/X, _and so for any sheaf F of abelian groups on X_

$$\check{H}^n(X, F) \cong H^n(X, F) \quad \text{for } n = 0, 1.$$

Proof. Any $f_1 \in \check{H}^0(X, S)$ is represented by some

$$f \in \ker(d_0 : C^0(\mathfrak{U}, S) \to C^1(\mathfrak{U}, S))$$

for some open cover $\mathfrak{U} = (U_i)_{i \in I}$ of X. That is, $f = (f_i)_{i \in I} \in \Pi_{i \in I} S(U_i)$. But S has stalk zero everywhere; hence for each i, U_i has an open cover $(V_j)_{j \in J_i}$ such that

$$\forall j \in J_i \quad \rho^{U_i}_{V_j}(f_i) = 0.$$

Then $\mathfrak{V} = (V_j)_{j \in \cup_i J_i}$ is an open cover of X, refining $\mathfrak{U}$, such that $f \mapsto 0$ under the refinement map

$$\check{H}^0(\mathfrak{U}, S) \to \check{H}^0(\mathfrak{V}, S). \quad //$$

4.15 **Theorem.** _If_ X _is a topological space and_ $\mathfrak{U}$, $\mathfrak{V}$ _are two open covers of_ X _with_ $\mathfrak{V}$ _a refinement of_ $\mathfrak{U}$, _then for any sheaf_ F _of abelian groups on_ X, _the refinement map_

$$H^1(\mathfrak{U}, F) \to H^1(\mathfrak{V}, F)$$

is injective. Hence $H^1(X, F) = \check{H}^1(X, F)$ _is a union of subgroups isomorphic to the_ $H^1(\mathfrak{U}, F)$.

Proof. Let $\mathfrak{U} = (U_i)_{i \in I}$, $\mathfrak{V} = (V_j)_{j \in J}$ and let $r : J \to I$ be a refinement map.

Given $s \in \ker(d_1 : C^1(\mathfrak{U}, F) \to C^2(\mathfrak{U}, F))$ such that $\exists f \in \operatorname{Im}(d_0 : C^0(\mathfrak{V}, F) \to C^1(\mathfrak{V}, F))$ with $\forall j, k \in J$

$$\rho^{U_{rj,rk}}_{V_{jk}} s(rj, rk) = \rho^{V_j}_{V_{jk}}(f(j)) - \rho^{V_k}_{V_{jk}}(f(k))$$

we wish to show that

$$\exists \bar{f} \in \operatorname{Im}(d_0 : C^0(\mathfrak{U}, F) \to C^1(\mathfrak{U}, F)) \text{ with } \forall i, l \in I$$

$$s(i, l) = \bar{f}(i) - \bar{f}(l);$$

here we are writing V_{jk}, $s(j, k)$ for V_σ, s_σ where σ is the 1-simplex

$0 \mapsto j$, $1 \mapsto k$ and similarly for 0-simplices.

Suppose we are given $i \in I$: to construct $\overline{f}(i) \in \Gamma(U_i, F)$ we consider for each $x \in U_i$, and for each $j \in J$ such that $x \in V_j$, the element

$$\overline{f}_j = \rho^{V_j}_{U_i \cap V_j}(f(j)) - \rho^{U_{rj,i}}_{U_i \cap V_j}(s(rj, i))$$
$$\in \Gamma(U_i \cap V_j, F).$$

(For the rest of the proof we omit the restriction maps and instead name the domain on which equations hold.)

If $k \in J$ is also such that $x \in V_k$, then

$$\begin{aligned}\overline{f}_j - \overline{f}_k &= f(j) - f(k) + s(rk, i) - s(rj, i)\\ &= s(rj, rk) + s(rk, i) - s(rj, i)\\ &= 0 \quad \text{on } U_i \cap V_j \cap V_k\end{aligned}$$

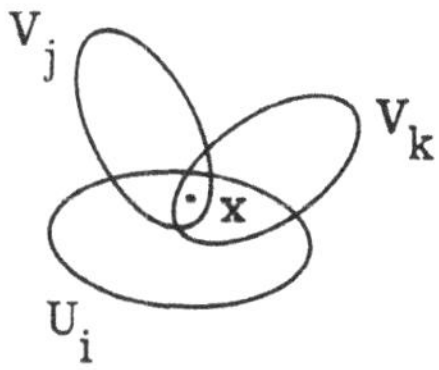

since $s \in \ker d_1$. Thus the $\overline{f}_j$ $(j \in J)$ glue to give an element $\overline{f}(i) \in \Gamma(U_i, F)$.

Furthermore, on $U_i \cap U_l$ $(i, l \in I)$ we have

$$s(i, l) = \overline{f}(i) - \overline{f}(l)$$

since $\forall x \in U_i \cap U_l$, whenever $x \in U_{il} \cap V_{jk}$ for some $j, k \in J$

$$\begin{aligned}\overline{f}(i) - \overline{f}(l) &= f(j) - s(rj, i) - f(k) + s(rk, l)\\ &= s(rj, rk) - s(rj, i) + s(rk, l)\\ &= -s(rk, i) + s(rk, l)\\ &= s(i, l) \quad \text{on } U_{il} \cap V_{jk}.\end{aligned}$$

The result follows, and the last part follows from the construction of the direct limit (cf. 1. Ex. 5). //

4.16 Čech cohomology has a connexion with the picard group (4.5.5 and 4.5.6), given as follows.

Let $(X, \mathcal{O})$ be a ringed space. An invertible $\mathcal{O}$-Module M is given by the following data:

(a) an open covering $\mathfrak{U} = (U_i)_{i \in I}$ of X (such that $\forall i \in I$ $M|U_i \cong \mathcal{O}|U_i$);

(b) for each $i, j \in I$ an isomorphism of $\mathcal{O}|(U_i \cap U_j)$-Modules

$$\mathcal{O}|(U_i \cap U_j) \xrightarrow{\sim} \mathcal{O}|(U_j \cap U_i)$$

(each is isomorphic to $M|(U_i \cap U_j)$).

By 4.5.3 the data of (b) is equivalent to giving, for each $(i, j) \in I \times I$, a unit

$$f_{ij} \in \Gamma(U_i \cap U_j, \mathcal{O})$$

(so that the isomorphism of (b) is 'multiply by f_{ij}').

Now the assignment $U \mapsto (\Gamma(U, \mathcal{O}))^* =$ group of units of $\Gamma(U, \mathcal{O})$ defines a sheaf $\mathcal{O}^*$ of abelian groups (under multiplication), and the f_{ij} give an element

$$f = (f_{ij})_{(i, j) \in I_1} \in C^1(\mathfrak{U}, \mathcal{O}^*).$$

Since the isomorphisms in (b) are compatible on the triple intersections $U_i \cap U_j \cap U_k$, f is in fact a cocycle, that is

$$f \in \ker(d_1 : C^1(\mathfrak{U}, \mathcal{O}^*) \to C^2(\mathfrak{U}, \mathcal{O}^*)) = Z^1(\mathfrak{U}, \mathcal{O}^*) \text{ say.}$$

Conversely, given $f \in Z^1(\mathfrak{U}, \mathcal{O}^*)$ we can construct an $\mathcal{O}$-Module M by glueing the copies of $\mathcal{O}|(U_i \cap U_j)$ by the recipe given by f in (b). Hence we have defined a map

$$\zeta : Z^1(\mathfrak{U}, \mathcal{O}^*) \to \text{Pic } X$$

which has as image the set of isomorphism classes of invertible sheaves which are trivialized by $\mathfrak{U}$ (that is, are such that $\forall i \in I \; M|U_i \cong \mathcal{O}|U_i$). It is also clear from the construction that ζ is a morphism of abelian groups, that is it takes the composition of cocycles (written naturally as multiplication in this case) to the operation on Pic X induced by $\otimes$.

Suppose now $f = (f_{ij})$ is a member of the kernel of ζ; then the invertible sheaf M constructed from f as above is trivial:

$$M \cong \mathcal{O}.$$

Now we have a global section $1 \in \Gamma(X, \mathcal{O}^*)$; let

$$g_i \in \Gamma(U_i, M)$$

be the corresponding section. Then by (b)

$$\forall i, j \in I \quad g_j = f_{ij} g_i \quad \text{on } U_i \cap U_j,$$

so f is a coboundary, that is

$$f \in \operatorname{Im}(d_0 : C^0(\mathfrak{U}, \mathcal{O}^*) \to C^1(\mathfrak{U}, \mathcal{O}^*)).$$

Hence ζ induces an injection

$$H^1(\mathfrak{U}, \mathcal{O}^*) \to \operatorname{Pic} X$$

with image that of ζ (cf. 4.14).

Since every invertible sheaf is trivial over some covering, and the refinement maps are easily seen to be compatible with the maps ζ, we obtain

Theorem. There is a natural isomorphism of abelian groups

$$H^1(X, \mathcal{O}^*) = \check{H}^1(X, \mathcal{O}^*) \cong \operatorname{Pic} X. \quad //$$

4.17 As an application, let $(X, \mathcal{O})$ be a complex manifold (continuous, differentiable or analytic). Then there is a sheaf morphism $\mathcal{O} \to \mathcal{O}^*$ which sends a $\mathbf{C}$-valued function f to the function $\exp(2\pi i f)$; this provides a short exact sequence of sheaves of abelian groups:

$$0 \to \mathbf{Z} \to \mathcal{O} \to \mathcal{O}^* \to 0$$

where $\mathbf{Z}$ denotes the constant sheaf (of integer-valued functions).

The associated cohomology sequence contains the map

$$\partial : \operatorname{Pic} X = H^1(X, \mathcal{O}^*) \to H^2(X, \mathbf{Z})$$

and by 4.13 the target group can be interpreted as the topological (singular) cohomology of X with integral coefficients. When reinterpreted as a map on invertible sheaves, ∂ is called the Chern class map.

Exercises on Chapter 5

1. Use 4.15 to show that $H^1(\mathbf{C}^*, \mathbf{C}) \neq \{0\}$, where $\mathbf{C}^* = \mathbf{C}\setminus\{0\}$, and $\mathbf{C}$ denotes the constant sheaf; this 'explains' 3.6.10.

2. Let $G : K \to K'$ be a functor between abelian categories; we say that G is effaceable iff for each object A of K we can find a monomorphism $u : A \to E$ in K such that $G(u) = 0$. A ∂-functor $\{G^n : 0 \leq n < a\}$ is called effaceable iff G^n is for $0 < n < a$.

Show that if K has enough injectives, this is equivalent to the definition given in 2.13.

Show that, for any abelian category K, an exact, effaceable (in this new sense) ∂-functor $F^{\cdot}$ over a left exact functor $F : K \to K'$ has the universal property of 2.13 (for $R^{\cdot} F$).

3. Let X be a topological space, considered with the structure sheaf $\mathcal{O} = \mathbf{Z}_X$, the constant sheaf $\mathbf{Z}$, so that $\mathcal{O}$-Mod = Shv/X is the category of sheaves of abelian groups on X. Let A be a closed subspace of X, with inclusion map $j : A \hookrightarrow X$.

(a) Let G be a sheaf of abelian groups on A and $G^X = j_*G$ its extension by zero to X. Show that there is a natural isomorphism

$$H^*(X, G^X) \cong H^*(A, G).$$

(b) Deduce that if F is a sheaf of abelian groups on X, there is an exact cohomology sequence

$$\ldots \to H^n(X, F_U) \to H^n(X, F) \to H^n(A, F|A) \to H^{n+1}(X, F_U) \to \ldots$$

where $U = X\setminus A$ and $F_U = (F|U)^X$.

4. Establish that for any topological space X and sheaf F of abelian groups on X there is a natural map

$$\check{H}^2(X, F) \to H^2(X, F)$$

which is always injective, in two ways:

(a) directly, rather as in 4.13 and 4.14

(b) by first constructing the spectral sequence

$$\check{H}^p(X, \mathcal{H}^q(F)) \Rightarrow H^{p+q}(X, F)$$

(where $\mathcal{H}^q(F)$ is the presheaf $U \mapsto H^q(U, F)$ mentioned in 4.13). [One way to do this is to obtain it as the spectral sequence of a composite functor ([T], 2.4.1), using 4.10; compare Artin, Grothendieck Topologies, II §3.]

5. Use the ideas of 4.5 and 4.12 (chain homotopies) to show that

$$H^*(\mathfrak{U}, F) = H^*(C_1^{\cdot}(\mathfrak{U}, F))$$

where $C_1^{\cdot}(\mathfrak{U}, F)$ is the subcomplex of $C^{\cdot}(\mathfrak{U}, F)$ consisting of the <u>alternating</u> cochains, namely those (s_σ) for which

(a) $s_\sigma = \varepsilon s_{\sigma'}$ if the values of σ' form a permutation of signature ε of the values of σ; and

(b) $s_\sigma = 0$ if σ takes two equal values.

Suppose that X admits a covering $\mathfrak{U} = (U_i)_{i \in I}$ such that for some $n \in \mathbf{N}$, $U_\sigma = \emptyset$ for all those $\sigma \in I_n$ which take distinct values (we then say $\dim X \leq n$). Deduce that for any sheaf F on X we have

$$m > n \Rightarrow \check{H}^m(X, F) = 0.$$

[Compare Serre, Faisceaux algébriques cohérents (Annals <u>61</u> (1955) 197-278) I. 18 and I. 20; and [G] II. 5.12.]

6. Suppose that $\mathfrak{U}$ is an open cover of X which is <u>acyclic</u> for a given sheaf F; that is, for each simplex σ,

$$r > 0 \Rightarrow H^r(U_\sigma, F|U_\sigma) = 0.$$

Show that

$$H^*(\mathfrak{U}, F) \cong H^*(X, F).$$

[This is hard; see [G] Cor du Th. 5.4.1, or Serre, FAC (see Q5) I. 29.]

The way ahead: further reading

The grounding in sheaf theory given by this course should enable the reader to proceed with the study of a number of subjects. Some suggestions are given below: there is no particular significance in the ordering.

(a) In topology, one can study cohomology operations (cup and cap products), the Leroy and Serre spectral sequences, and Borel-Moore homology. See for example G. E. Bredon, Sheaf Theory (McGraw-Hill 1967); R. G. Swan, The Theory of Sheaves (Chicago U. P. 1964); and [G].

(b) In algebraic geometry, only the basic machinery has been indicated. One must now do more work on affine schemes to understand the local nature of an algebraic variety, and use sheaf theory to connect this with the global properties. A very good introduction is D. B. Mumford, Introduction to Algebraic Geometry (mimeographed notes from Harvard). See also I. G. Macdonald, Algebraic Geometry (Benjamin 1968); and [EGA I-IV], although this cannot be recommended whole-heartedly as reading material.

For specific examples of the applications of sheaf theory, there is the excellent paper J-P. Serre, Faisceaux algébriques cohérents (Annals, 61 (1955) 197-278), which uses an older definition of algebraic variety; D. B. Mumford, Lectures on Curves on an Algebraic Surface (Princeton U. P., 1966), which relies heavily on cohomology, and incidentally gives a rapid outline introduction to scheme theory; and Y. I. Manin, Lectures on the K-functor in Algebraic Geometry (Russian Math. Surveys, 24 (1969), No. 5, pp. 1-90).

(c) The applications of sheaf-theoretic topology to (classical) algebraic geometry (such as the various generalisations of the Riemann-Roch Theorem) are well represented in [H]; see also Atiyah, K-Theory (Benjamin, 1967).

(d) In working out the abstract machinery necessary to attack the Weil conjectures, Grothendieck and his school were led to a generalisation of topological spaces, over which one can still do sheaf theory: these are the Grothendieck topologies. A good introduction, showing how neat and ultimately categorical sheaf theory can be made is in M. Artin, Grothendieck Topologies (Harvard Lecture Notes, 1962); see also Mumford's chapter in Arithmetical Algebraic Geometry (edited by Schilling; Harper and Row, 1965). The bible (or elbib?) of this sect is SGA4 (Springer Lecture Notes 269, 270 and 305); it is by no means easy reading. See also H. Schubert, Categories, Chapter 20 (Springer, 1972); M. Hakim, Topos annelés et schémas relatifs (Springer, 1972).

(e) The latest abstraction from (d) is the theory of elementary topoi; it is an ambitious attempt to unite geometry and set theory, and is at an early stage of development. See A. Kock and G. C. Wraith, Elementary Toposes (Aarhus Lecture Notes No. 30, 1971); P. Freyd, Aspects of Topoi (Bull. Aust. Math. Soc., 7 (1972) 1-76); F. W. Lawvere, Quantifiers and Sheaves (Proceedings of the International Congress of Mathematicians, Nice 1970; Vol. I, pp. 329-34).

References

[EGA I] A. Grothendieck and J. A. Dieudonné, Eléments de géométrie algébrique I (Second Edition); Springer, 1971.

[EGA II, III] A. Grothendieck and J. A. Dieudonné, Eléments de géométrie algébrique II, III; IHES Publ. Math. 8, 11, 17.

[G] R. Godement, Topologie algébrique et théorie des faisceaux; Hermann, 1964.

[H] F. Hirzebruch, Topological methods in algebraic geometry (Third Edition); Springer, 1966.

[K] J. T. Knight, Commutative algebra; Cambridge University Press, 1971.

[L] S. Lang, Differential manifolds; Addison-Wesley, 1972.

[Macl] S. Maclane, Categories for the working mathematician; Springer, 1971.

[Mit] B. Mitchell, Theory of categories; Academic Press, 1965.

[T] A. Grothendieck, Sur quelques points d'algèbre homologique; Tohoku Math. Journal, IX (1957), pp. 119-221.

Hints and answers to some exercises

1\. Ex. 3(iii) No: only torsion groups are so obtainable.

1\. Ex. 5 All the maps to the direct limit are injective (respectively surjective).

2\. 1. 9 Use the empty cover of the empty set. Aliter, apply 2. 3. 1 (or its proof).

2\. Ex. 1 $\Gamma(I, F) = \mathbf{Z} \oplus \mathbf{Z}$; just one; as many as there are pairs of abelian group morphisms, each $\mathbf{Z} \to \mathbf{Z}$.

3\. 6. 2 See 4. 1. 13.

3\. 6. 8 $F \to G \to H$ is exact iff for some objects K, L and morphisms $F \to K \to G \to L \to H$ the three sequences

$$F \to K \to 0,$$
$$0 \to K \to G \to L \to 0,$$
and
$$0 \to L \to H$$

are exact. This condition is preserved under an exact T.

3\. Ex. 7 First show that a short exact sequence $0 \to P \to Q \overset{s}{\to} R \to 0$ is split (that is $\exists t : R \to Q$ such that $s \circ t = id_R$) iff $Q \cong P \oplus R$ with s corresponding to the natural projection. Hence (a) $\Rightarrow$ (b). Now for $s, t : P \to P$, factorise $(s - t)$ as

$$P \to P \oplus P \overset{(s,t)}{\to} P \oplus P \to P.$$

Hence (b) $\Rightarrow$ (c). For (c) $\Rightarrow$ (b) show that $X \cong P \oplus Q$ iff there are morphisms

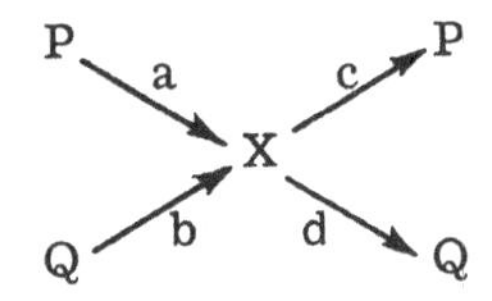

such that $ca = id_P$, $db = id_Q$, $da = 0$, $cb = 0$ and $ac + bd = id_X$.

4.3.5 Spec S → Spec R induces a morphism of global sections

$$R = \Gamma(\text{Spec } R, \mathcal{O}_R) \to \Gamma(\text{Spec } S, \mathcal{O}_S) = S.$$

Compare 4.3.11.

5. Ex. 3 (a) Use 3.8.8(a) to see that $H^*(X, -^X)$ is an exact ∂-functor over $\Gamma(A, -)$, and 3.7.13 to show that it is effaceable; apply 2.13.

(b) Use 3.8.11.

Index of terminology

Index of notation